*Flavors of Geometry* is a volume of lectures on four geometrically influenced fields of mathematics that have experienced great development in recent years.

Growing out of a series of introductory lectures given at the Mathematical Sciences Research Institute in January 1995 and January 1996, the book presents chapters on hyperbolic geometry, dynamics in several complex variables, convex geometry, and volume estimation, by masters in their respective fields. Each lecture begins with a discussion of elementary concepts, examines the highlights of the field, and concludes with a look at more advanced material. The style and presentation of the chapters are clear and accessible, and most of the lectures are richly illustrated. Bibiliographies and indexes are included to encourage further reading on the topics discussed.

These lectures are an excellent starting place for graduate students and will also interest researchers wanting a "flavor" of new developments in geometry.

Mathematical Sciences Research Institute
Publications

31

Flavors of Geometry

# Mathematical Sciences Research Institute
# Publications

**Volumes 1 through 27 are available from Springer-Verlag**

# Flavors of Geometry

*Edited by*

**Silvio Levy**

*Mathematical Sciences Research Institute*

Silvio Levy
Mathematical Sciences Research
Institute
1000 Centennial Drive
Berkeley, CA 94720
levy@math.berkeley.edu

Mathematical Sciences Research
Institute
1000 Centennial Drive
Berkeley, CA 94720

The Mathematical Sciences Research Institute wishes to acknowledge
support by the National Science Foundation.

---

CAMBRIDGE UNIVERSITY PRESS
Cambridge, New York, Melbourne, Madrid, Cape Town, Singapore, São Paulo

Cambridge University Press
The Edinburgh Building, Cambridge CB2 8RU, UK

Published in the United States of America by Cambridge University Press, New York

www.cambridge.org
Information on this title: www.cambridge.org/9780521620482

Flavors of Geometry
MSRI Publications
Volume **31**, 1997

# Contents

Flavors of Geometry
MSRI Publications
Volume **31**, 1997

# Preface

> ... βούλεσθ', εἶπεν, ἐπεὶ λόγοι περὶ θεῶν γεγόνασιν,
> ἐν τοῖς Πλάτωνος γενεθλίοις αὐτὸν Πλάτωνα κοινωνὸν
> παραλάβωμεν, ἐπισκεψάμενοι τίνα λαβὼν γνώμην
> ἀπεφήνατ' ἀεὶ γεωμετρεῖν τὸν θεόν; εἴ γε δὴ θετέον εἶναι
> τὴν ἀπόφασιν ταύτην Πλάτωνος.
>
> Ἐμοῦ δὲ ταῦτ' εἰπόντος ὡς γέγραπται μὲν ἐν οὐδενὶ
> σαφῶς τῶν ἐκείνου βυβλίων, ἔχει δὲ πίστιν ἱκανὴν καὶ
> τοῦ πλατωνικοῦ χαρακτῆρός ἐστιν ...
>
> — Plutarch, *Quaest. Conv.* VIII.2, in his *Moralia*

*God is always doing geometry.* What Plato meant by this is no clearer now than it was in the first century A.D., when the dinner conversation quoted above (which I'm about to paraphrase) took place. Diogenianus, one of the guests, recalls that it is Plato's birthday and proposes to "bring him into the conversation" by debating the meaning of this nugget—"if indeed the statement is Plato's." Plutarch, who knew a lot about such things, replies that although the quote does not appear explicitly in any of Plato's books, it is well enough attested, and it is in character.

There follow the opinions of the three remaining guests, and finally that of Plutarch. Drastically abridged, they are:

- Geometry focuses the mind on the abstract rather than on the sensorial.
- Geometry is divine in that "geometric proportion ... befits a moderate oligarchy or a lawful monarchy ... it distributes to each according to his worth," whereas arithmetic proportion is egalitarian.
- The essence of geometry is boundaries, so God does geometry when he bounds matter to create.
- Geometry intervenes when proportion and measure and number are used to order chaotic nature.

Although encompassing, these are not entirely persuasive explanations; make of them what you will. I bring them up largely for fun, but also to illustrate the breadth of interpretation that the notion of geometry has traditionally enjoyed, and so to defend my choice of title for this volume. One might consider also Klein's definition in his Erlangen program. And, while the study of complex dynamics in one or more variables involves tools from analysis, topology, and so on, a large part of the motivation is geometric: we want to understand the qualitative and spatial behavior of orbits.

*       *       *

This book is based on introductory lecture series that took place at MSRI in 1995 and 1996, aimed at graduate students and mathematicians in all fields.

The first series of lectures ran from January 9 to 20, 1995, as a prelude to the Spring 1995 program in Complex Dynamics and Hyperbolic Geometry. It consisted of three courses of five lectures each:

- *Hyperbolic Geometry* by James Cannon;
- *Conformal Dynamics on the Riemann Sphere* by John Hubbard;
- *Complex Dynamics in Several Variables* by John Smillie.

Of these, the first and last are included in this volume; notes for Smillie's lectures (pp. 117–150) were written by Gregery Buzzard, and those for Cannon's lectures (pp. 59–115) primarily by him, with the help of Bill Floyd, Richard Kenyon, and Walter Parry.

The second series of lectures ran from January 29 to February 9, 1996, in conjunction with the Spring 1996 program on Convex Geometry and Geometric Functional Analysis. Four courses of 3–4 lectures each were offered:

- *Classical Convex Geometry* by Keith Ball;
- *Concentration of Measure in Geometry* by Gideon Schechtman;
- *Spherical Sections of Octahedra* by Joram Lindenstrauss;
- *Random Methods for Volume Computation* by Béla Bollobás.

Keith Ball kindly agreed to merge the first three courses into one set of notes (pp. 1–58), so that Schechtman's and Lindenstrauss's material is now, after considerable transmutation, integrated with the more basic material of Ball's lectures. Bollobás wrote up his own lectures (pp. 151–182).

I've arranged the four contributions from easier to harder, but relative difficulty depends somewhat on one's background, so you should browse. Each set of notes starts with elementary concepts, proceeds through highlights of the field, and concludes with a taste of advanced material. Some math undergraduates will find most of Ball's and Cannon's contributions, and at least the beginning of the others, perfectly accessible.

For ease of reference I have supplied an index for each set of notes. Any shortcomings the indexes may have are to be blamed on me, not on the authors.

Silvio Levy<br>Berkeley, February 1997

Flavors of Geometry
MSRI Publications
Volume **31**, 1997

# Note on MSRI Programs

The introductory courses and workshops that gave rise to this book were part of the plan, advanced by the MSRI directorate during the last several years, of intensifying the Institute's effectiveness and outreach by means other than research in core mathematics (which nonetheless remains the center of our activities). Some others among these outreach efforts:

- Conversations between Researchers and Teachers, a series of presentations and discussions bringing together two groups that rarely mix: high-school math teachers and research mathematicians.
- A conference on math visualization, one on The Future of Mathematical Communication, and one on The Future of Mathematical Education at Research Universities.
- The Fermat Fest, which explained to a lay public of 1000 the meaning and basic ideas of Fermat's Last Theorem.
- Numbers in Action, another event for the general public focusing on number theory.
- A program for broadcasting talks on the MBone (the Internet multicast backbone) and for helping other sites obtain access to the MBone.
- The Conference for African American Researchers and the Julia Robinson Celebration of Women in Mathematics Conference.
- Research workshops and programs on nontraditional topics, such as financial mathematics and combinatorial game theory.

Details about these events and programs, as well as most other MSRI activities, can be found at http://www.msri.org.

Flavors of Geometry
MSRI Publications
Volume **31**, 1997

# An Elementary Introduction
# to Modern Convex Geometry

### KEITH BALL

#### CONTENTS

## Preface

These notes are based, somewhat loosely, on three series of lectures given by myself, J. Lindenstrauss and G. Schechtman, during the Introductory Workshop in Convex Geometry held at the Mathematical Sciences Research Institute in Berkeley, early in 1996. A fourth series was given by B. Bollobás, on rapid mixing and random volume algorithms; they are found elsewhere in this book.

The material discussed in these notes is not, for the most part, very new, but the presentation has been strongly influenced by recent developments: among other things, it has been possible to simplify many of the arguments in the light of later discoveries. Instead of giving a comprehensive description of the state of the art, I have tried to describe two or three of the more important ideas that have shaped the modern view of convex geometry, and to make them as accessible

as possible to a broad audience. In most places, I have adopted an informal style
that I hope retains some of the spontaneity of the original lectures. Needless to
say, my fellow lecturers cannot be held responsible for any shortcomings of this
presentation.

I should mention that there are large areas of research that fall under the
very general name of convex geometry, but that will barely be touched upon in
these notes. The most obvious such area is the classical or "Brunn–Minkowski"
theory, which is well covered in [Schneider 1993]. Another noticeable omission is
the combinatorial theory of polytopes: a standard reference here is [Brøndsted
1983].

## Lecture 1. Basic Notions

The topic of these notes is convex geometry. The objects of study are convex bodies: compact, convex subsets of Euclidean spaces, that have nonempty
interior. Convex sets occur naturally in many areas of mathematics: linear programming, probability theory, functional analysis, partial differential equations,
information theory, and the geometry of numbers, to name a few.

Although convexity is a simple property to formulate, convex bodies possess
a surprisingly rich structure. There are several themes running through these
notes: perhaps the most obvious one can be summed up in the sentence: "All
convex bodies behave a bit like Euclidean balls." Before we look at some ways in
which this is true it is a good idea to point out ways in which it definitely is not.
This lecture will be devoted to the introduction of a few basic examples that we
need to keep at the backs of our minds, and one or two well known principles.

The only notational conventions that are worth specifying at this point are
the following. We will use $|\cdot|$ to denote the standard Euclidean norm on $\mathbb{R}^n$. For
a body $K$, $\mathrm{vol}(K)$ will mean the volume measure of the appropriate dimension.

The most fundamental principle in convexity is the *Hahn–Banach separation
theorem*, which guarantees that each convex body is an intersection of half-spaces,
and that at each point of the boundary of a convex body, there is at least one
supporting hyperplane. More generally, if $K$ and $L$ are disjoint, compact, convex
subsets of $\mathbb{R}^n$, then there is a linear functional $\phi : \mathbb{R}^n \to \mathbb{R}$ for which $\phi(x) < \phi(y)$
whenever $x \in K$ and $y \in L$.

The simplest example of a convex body in $\mathbb{R}^n$ is the cube, $[-1, 1]^n$. This does
not look much like the Euclidean ball. The largest ball inside the cube has radius
1, while the smallest ball containing it has radius $\sqrt{n}$, since the corners of the
cube are this far from the origin. So, as the dimension grows, the cube resembles
a ball less and less.

The second example to which we shall refer is the $n$-dimensional regular solid
simplex: the convex hull of $n + 1$ equally spaced points. For this body, the ratio
of the radii of inscribed and circumscribed balls is $n$: even worse than for the
cube. The two-dimensional case is shown in Figure 1. In Lecture 3 we shall see

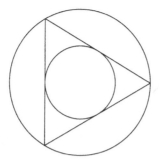

**Figure 1.** Inscribed and circumscribed spheres for an $n$-simplex.

that these ratios are extremal in a certain well-defined sense.

Solid simplices are particular examples of cones. By a *cone* in $\mathbb{R}^n$ we just mean the convex hull of a single point and some convex body of dimension $n-1$ (Figure 2). In $\mathbb{R}^n$, the volume of a cone of height $h$ over a base of $(n-1)$-dimensional volume $B$ is $Bh/n$.

The third example, which we shall investigate more closely in Lecture 4, is the $n$-dimensional "octahedron", or *cross-polytope:* the convex hull of the $2n$ points $(\pm 1, 0, 0, \ldots, 0)$, $(0, \pm 1, 0, \ldots, 0)$, $\ldots$, $(0, 0, \ldots, 0, \pm 1)$. Since this is the unit ball of the $\ell_1$ norm on $\mathbb{R}^n$, we shall denote it $B_1^n$. The circumscribing sphere of $B_1^n$ has radius 1, the inscribed sphere has radius $1/\sqrt{n}$; so, as for the cube, the ratio is $\sqrt{n}$: see Figure 3, left.

$B_1^n$ is made up of $2^n$ pieces similar to the piece whose points have nonnegative coordinates, illustrated in Figure 3, right. This piece is a cone of height 1 over a base, which is the analogous piece in $\mathbb{R}^{n-1}$. By induction, its volume is

$$\frac{1}{n} \cdot \frac{1}{n-1} \cdot \cdots \cdot \frac{1}{2} \cdot 1 = \frac{1}{n!},$$

and hence the volume of $B_1^n$ is $2^n/n!$.

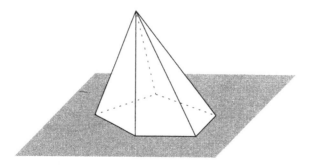

**Figure 2.** A cone.

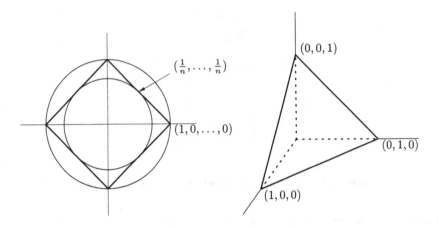

**Figure 3.** The cross-polytope (left) and one orthant thereof (right).

The final example is the Euclidean ball itself,

$$B_2^n = \left\{ x \in \mathbb{R}^n : \sum_1^n x_i^2 \leq 1 \right\}.$$

We shall need to know the volume of the ball: call it $v_n$. We can calculate the surface "area" of $B_2^n$ very easily in terms of $v_n$: the argument goes back to the ancients. We think of the ball as being built of thin cones of height 1: see Figure 4, left. Since the volume of each of these cones is $1/n$ times its base area, the surface of the ball has area $nv_n$. The sphere of radius 1, which is the surface of the ball, we shall denote $S^{n-1}$.

To calculate $v_n$, we use integration in spherical polar coordinates. To specify a point $x$ we use two coordinates: $r$, its distance from 0, and $\theta$, a point on the sphere, which specifies the direction of $x$. The point $\theta$ plays the role of $n-1$ real coordinates. Clearly, in this representation, $x = r\theta$: see Figure 4, right. We can

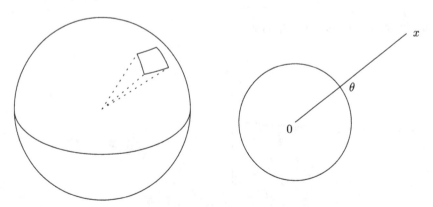

**Figure 4.** Computing the volume of the Euclidean ball.

write the integral of a function on $\mathbb{R}^n$ as

$$\int_{\mathbb{R}^n} f = \int_{r=0}^{\infty} \int_{S^{n-1}} f(r\theta) \text{ “}d\theta\text{” } r^{n-1} \, dr. \tag{1.1}$$

The factor $r^{n-1}$ appears because the sphere of radius $r$ has area $r^{n-1}$ times that of $S^{n-1}$. The notation "$d\theta$" stands for "area" measure on the sphere: its total mass is the surface area $nv_n$. The most important feature of this measure is its rotational invariance: if $A$ is a subset of the sphere and $U$ is an orthogonal transformation of $\mathbb{R}^n$, then $UA$ has the same measure as $A$. Throughout these lectures we shall normalise integrals like that in (1.1) by pulling out the factor $nv_n$, and write

$$\int_{\mathbb{R}^n} f = nv_n \int_0^{\infty} \int_{S^{n-1}} f(r\theta) r^{n-1} \, d\sigma(\theta) \, dr$$

where $\sigma = \sigma_{n-1}$ is the rotation-invariant measure on $S^{n-1}$ of total mass 1. To find $v_n$, we integrate the function

$$x \mapsto \exp\left(-\tfrac{1}{2}\sum_1^n x_i^2\right)$$

both ways. This function is at once invariant under rotations and a product of functions depending upon separate coordinates; this is what makes the method work. The integral is

$$\int_{\mathbb{R}^n} f = \int_{\mathbb{R}^n} \prod_1^n e^{-x_i^2/2} \, dx = \prod_1^n \left(\int_{-\infty}^{\infty} e^{-x_i^2/2} \, dx_i\right) = (\sqrt{2\pi})^n.$$

But this equals

$$nv_n \int_0^{\infty} \int_{S^{n-1}} e^{-r^2/2} r^{n-1} \, d\sigma \, dr = nv_n \int_0^{\infty} e^{-r^2/2} r^{n-1} \, dr = v_n 2^{n/2} \Gamma\left(\frac{n}{2}+1\right).$$

Hence

$$v_n = \frac{\pi^{n/2}}{\Gamma\left(\frac{n}{2}+1\right)}.$$

This is extremely small if $n$ is large. From Stirling's formula we know that

$$\Gamma\left(\frac{n}{2}+1\right) \sim \sqrt{2\pi} \, e^{-n/2} \left(\frac{n}{2}\right)^{(n+1)/2},$$

so that $v_n$ is roughly

$$\left(\sqrt{\frac{2\pi e}{n}}\right)^n.$$

To put it another way, the Euclidean ball of *volume* 1 has *radius* about

$$\sqrt{\frac{n}{2\pi e}},$$

which is pretty big.

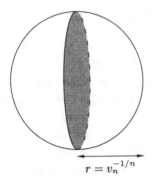

$$r = v_n^{-1/n}$$

**Figure 5.** Comparing the volume of a ball with that of its central slice.

This rather surprising property of the ball in high-dimensional spaces is perhaps the first hint that our intuition might lead us astray. The next hint is provided by an answer to the following rather vague question: how is the mass of the ball distributed? To begin with, let's estimate the $(n-1)$-dimensional volume of a slice through the centre of the ball of volume 1. The ball has radius

$$r = v_n^{-1/n}$$

(Figure 5). The slice is an $(n-1)$-dimensional ball of this radius, so its volume is

$$v_{n-1}r^{n-1} = v_{n-1}\left(\frac{1}{v_n}\right)^{(n-1)/n}.$$

By Stirling's formula again, we find that the slice has volume about $\sqrt{e}$ when $n$ is large. What are the $(n-1)$-dimensional volumes of parallel slices? The slice at distance $x$ from the centre is an $(n-1)$-dimensional ball whose radius is $\sqrt{r^2 - x^2}$ (whereas the central slice had radius $r$), so the volume of the smaller slice is about

$$\sqrt{e}\left(\frac{\sqrt{r^2 - x^2}}{r}\right)^{n-1} = \sqrt{e}\left(1 - \frac{x^2}{r^2}\right)^{(n-1)/2}.$$

Since $r$ is roughly $\sqrt{n/(2\pi e)}$, this is about

$$\sqrt{e}\left(1 - \frac{2\pi e x^2}{n}\right)^{(n-1)/2} \approx \sqrt{e}\exp(-\pi e x^2).$$

Thus, if we project the mass distribution of the ball of volume 1 onto a single direction, we get a distribution that is approximately Gaussian (normal) with variance $1/(2\pi e)$. What is remarkable about this is that the variance does not depend upon $n$. Despite the fact that the radius of the ball of volume 1 grows like $\sqrt{n/(2\pi e)}$, almost all of this volume stays within a slab of fixed width: for example, about 96% of the volume lies in the slab

$$\{x \in \mathbb{R}^n : -\tfrac{1}{2} \le x_1 \le \tfrac{1}{2}\}.$$

See Figure 6.

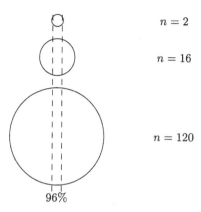

$n = 2$

$n = 16$

$n = 120$

96%

**Figure 6.** Balls in various dimensions, and the slab that contains about 96% of each of them.

So the volume of the ball concentrates close to *any* subspace of dimension $n - 1$. This would seem to suggest that the volume concentrates near the centre of the ball, where the subspaces all meet. But, on the contrary, it is easy to see that, if $n$ is large, most of the volume of the ball lies near its surface. In objects of high dimension, measure tends to concentrate in places that our low-dimensional intuition considers small. A considerable extension of this curious phenomenon will be exploited in Lectures 8 and 9.

To finish this lecture, let's write down a formula for the volume of a general body in spherical polar coordinates. Let $K$ be such a body with 0 in its interior, and for each direction $\theta \in S^{n-1}$ let $r(\theta)$ be the radius of $K$ in this direction. Then the volume of $K$ is

$$nv_n \int_{S^{n-1}} \int_0^{r(\theta)} s^{n-1} \, ds \, d\sigma = v_n \int_{S^{n-1}} r(\theta)^n \, d\sigma(\theta).$$

This tells us a bit about particular bodies. For example, if $K$ is the cube $[-1, 1]^n$, whose volume is $2^n$, the radius satisfies

$$\int_{S^{n-1}} r(\theta)^n = \frac{2^n}{v_n} \approx \left( \sqrt{\frac{2n}{\pi e}} \right)^n.$$

So the "average" radius of the cube is about

$$\sqrt{\frac{2n}{\pi e}}.$$

This indicates that the volume of the cube tends to lie in its corners, where the radius is close to $\sqrt{n}$, not in the middle of its facets, where the radius is close to 1. In Lecture 4 we shall see that the reverse happens for $B_1^n$ and that this has a surprising consequence.

If $K$ is (*centrally*) *symmetric*, that is, if $-x \in K$ whenever $x \in K$, then $K$ is the unit ball of some norm $\|\cdot\|_K$ on $\mathbb{R}^n$:

$$K = \{x : \|x\|_K \leq 1\}.$$

This was already mentioned for the octahedron, which is the unit ball of the $\ell_1$ norm

$$\|x\| = \sum_1^n |x_i|.$$

The norm and radius are easily seen to be related by

$$r(\theta) = \frac{1}{\|\theta\|}, \quad \text{for } \theta \in S^{n-1},$$

since $r(\theta)$ is the largest number $r$ for which $r\theta \in K$. Thus, for a general symmetric body $K$ with associated norm $\|\cdot\|$, we have this formula for the volume:

$$\text{vol}(K) = v_n \int_{S^{n-1}} \|\theta\|^{-n} \, d\sigma(\theta).$$

## Lecture 2. Spherical Sections of the Cube

In the first lecture it was explained that the cube is rather unlike a Euclidean ball in $\mathbb{R}^n$: the cube $[-1,1]^n$ includes a ball of radius 1 and no more, and is included in a ball of radius $\sqrt{n}$ and no less. The cube is a bad approximation to the Euclidean ball. In this lecture we shall take this point a bit further. A body like the cube, which is bounded by a finite number of flat facets, is called a *polytope*. Among symmetric polytopes, the cube has the fewest possible facets, namely $2n$. The question we shall address here is this:

*If $K$ is a polytope in $\mathbb{R}^n$ with $m$ facets, how well can $K$ approximate the Euclidean ball?*

Let's begin by clarifying the notion of approximation. To simplify matters we shall only consider symmetric bodies. By analogy with the remarks above, we could define the distance between two convex bodies $K$ and $L$ to be the smallest $d$ for which there is a scaled copy of $L$ inside $K$ and another copy of $L$, $d$ times as large, containing $K$. However, for most purposes, it will be more convenient to have an affine-invariant notion of distance: for example we want to regard all parallelograms as the same. Therefore:

DEFINITION. The distance $d(K, L)$ between symmetric convex bodies $K$ and $L$ is the least positive $d$ for which there is a linear image $\tilde{L}$ of $L$ such that $\tilde{L} \subset K \subset d\tilde{L}$. (See Figure 7.)

Note that this distance is multiplicative, not additive: in order to get a metric (on the set of linear equivalence classes of symmetric convex bodies) we would need to take $\log d$ instead of $d$. In particular, if $K$ and $L$ are identical then $d(K, L) = 1$.

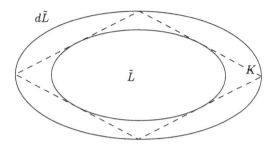

**Figure 7.** Defining the distance between $K$ and $L$.

Our observations of the last lecture show that the distance between the cube and the Euclidean ball in $\mathbb{R}^n$ is *at most* $\sqrt{n}$. It is intuitively clear that it really is $\sqrt{n}$, i.e., that we cannot find a linear image of the ball that sandwiches the cube any better than the obvious one. A formal proof will be immediate after the next lecture.

The main result of this lecture will imply that, if a polytope is to have small distance from the Euclidean ball, it must have very many facets: exponentially many in the dimension $n$.

THEOREM 2.1. *Let $K$ be a (symmetric) polytope in $\mathbb{R}^n$ with $d(K, B_2^n) = d$. Then $K$ has at least $e^{n/(2d^2)}$ facets. On the other hand, for each $n$, there is a polytope with $4^n$ facets whose distance from the ball is at most 2.*

The arguments in this lecture, including the result just stated, go back to the early days of packing and covering problems. A classical reference for the subject is [Rogers 1964].

Before we embark upon a proof of Theorem 2.1, let's look at a reformulation that will motivate several more sophisticated results later on. A symmetric convex body in $\mathbb{R}^n$ with $m$ pairs of facets can be realised as an $n$-dimensional slice (through the centre) of the cube in $\mathbb{R}^m$. This is because such a body is the intersection of $m$ slabs in $\mathbb{R}^n$, each of the form $\{x : |\langle x, v \rangle| \le 1\}$, for some nonzero vector $v$ in $\mathbb{R}^n$. This is shown in Figure 8.

Thus $K$ is the set $\{x : |\langle x, v_i \rangle| \le 1$ for $1 \le i \le m\}$, for some sequence $(v_i)_1^m$ of vectors in $\mathbb{R}^n$. The linear map

$$T : x \mapsto (\langle x, v_1 \rangle, \ldots, \langle x, v_m \rangle)$$

embeds $\mathbb{R}^n$ as a subspace $H$ of $\mathbb{R}^m$, and the intersection of $H$ with the cube $[-1, 1]^m$ is the set of points $y$ in $H$ for which $|y_i| \le 1$ for each coordinate $i$. So this intersection is the image of $K$ under $T$. Conversely, any $n$-dimensional slice of $[-1, 1]^m$ is a body with at most $m$ pairs of faces. Thus, the result we are aiming at can be formulated as follows:

*The cube in $\mathbb{R}^m$ has almost spherical sections whose dimension $n$ is roughly $\log m$ and not more.*

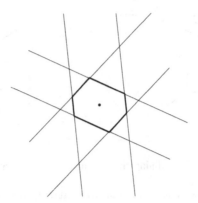

**Figure 8.** Any symmetric polytope is a section of a cube.

In Lecture 9 we shall see that all symmetric $m$-dimensional convex bodies have almost spherical sections of dimension about $\log m$. As one might expect, this is a great deal more difficult to prove for general bodies than just for the cube.

For the proof of Theorem 2.1, let's forget the symmetry assumption again and just ask for a polytope

$$K = \{x : \langle x, v_i \rangle \leq 1 \text{ for } 1 \leq i \leq m\}$$

with $m$ facets for which

$$B_2^n \subset K \subset dB_2^n.$$

What do these inclusions say about the vectors $(v_i)$? The first implies that each $v_i$ has length at most 1, since, if not, $v_i/|v_i|$ would be a vector in $B_2^n$ but not in $K$. The second inclusion says that if $x$ does not belong to $dB_2^n$ then $x$ does not belong to $K$: that is, if $|x| > d$, there is an $i$ for which $\langle x, v_i \rangle > 1$. This is equivalent to the assertion that for every unit vector $\theta$ there is an $i$ for which

$$\langle \theta, v_i \rangle \geq \frac{1}{d}.$$

Thus our problem is to look for as few vectors as possible, $v_1, v_2, \ldots, v_m$, of length at most 1, with the property that for every unit vector $\theta$ there is some $v_i$ with $\langle \theta, v_i \rangle \geq 1/d$. It is clear that we cannot do better than to look for vectors of length exactly 1: that is, that we may assume that all facets of our polytope touch the ball. Henceforth we shall discuss only such vectors.

For a fixed unit vector $v$ and $\varepsilon \in [0, 1)$, the set $C(\varepsilon, v)$ of $\theta \in S^{n-1}$ for which $\langle \theta, v \rangle \geq \varepsilon$ is called a *spherical cap* (or just a *cap*); when we want to be precise, we will call it the *$\varepsilon$-cap about $v$*. (Note that $\varepsilon$ does not refer to the radius!) See Figure 9, left.

We want every $\theta \in S^{n-1}$ to belong to at least one of the $(1/d)$-caps determined by the $(v_i)$. So our task is to estimate the number of caps of a given size needed to cover the entire sphere. The principal tool for doing this will be upper and lower estimates for the area of a spherical cap. As in the last lecture, we shall

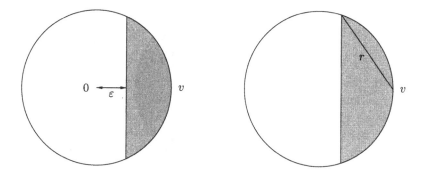

**Figure 9.** Left: $\varepsilon$-cap $C(\varepsilon, v)$ about $v$. Right: cap of radius $r$ about $v$.

measure this area as a proportion of the sphere: that is, we shall use $\sigma_{n-1}$ as our measure. Clearly, if we show that each cap occupies only a small proportion of the sphere, we can conclude that we need plenty of caps to cover the sphere. What is slightly more surprising is that once we have shown that spherical caps are not *too* small, we will also be able to deduce that we *can* cover the sphere without using too many.

In order to state the estimates for the areas of caps, it will sometimes be convenient to measure the size of a cap in terms of its radius, instead of using the $\varepsilon$ measure. The cap of radius $r$ about $v$ is

$$\left\{\theta \in S^{n-1} : |\theta - v| \leq r\right\}$$

as illustrated in Figure 9, right. (In defining the radius of a cap in this way we are implicitly adopting a particular metric on the sphere: the metric induced by the usual Euclidean norm on $\mathbb{R}^n$.) The two estimates we shall use are given in the following lemmas.

LEMMA 2.2 (UPPER BOUND FOR SPHERICAL CAPS). *For $0 \leq \varepsilon < 1$, the cap $C(\varepsilon, u)$ on $S^{n-1}$ has measure at most $e^{-n\varepsilon^2/2}$.*

LEMMA 2.3 (LOWER BOUND FOR SPHERICAL CAPS). *For $0 \leq r \leq 2$, a cap of radius $r$ on $S^{n-1}$ has measure at least $\frac{1}{2}(r/2)^{n-1}$.*

We can now prove Theorem 2.1.

PROOF. Lemma 2.2 implies the first assertion of Theorem 2.1 immediately. If $K$ is a polytope in $\mathbb{R}^n$ with $m$ facets and if $B_2^n \subset K \subset dB_2^n$, we can find $m$ caps $C(\frac{1}{d}, v_i)$ covering $S^{n-1}$. Each covers at most $e^{-n/(2d^2)}$ of the sphere, so

$$m \geq \exp\left(\frac{n}{2d^2}\right).$$

To get the second assertion of Theorem 2.1 from Lemma 2.3 we need a little more argument. It suffices to find $m = 4^n$ points $v_1, v_2, \ldots, v_m$ on the sphere so that the caps of radius 1 centred at these points cover the sphere: see Figure 10. Such a set of points is called a *1-net* for the sphere: every point of the sphere is

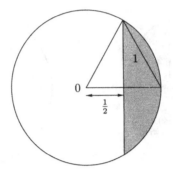

**Figure 10.** The $\frac{1}{2}$-cap has radius 1.

within distance 1 of some $v_i$. Now, if we choose a set of points on the sphere any two of which are at *least* distance 1 apart, this set cannot have too many points. (Such a set is called 1-*separated*.) The reason is that the caps of radius $\frac{1}{2}$ about these points will be disjoint, so the sum of their areas will be at most 1. A cap of radius $\frac{1}{2}$ has area at least $\left(\frac{1}{4}\right)^n$, so the number $m$ of these caps satisfies $m \le 4^n$. This does the job, because a *maximal* 1-separated set is automatically a 1-net: if we can't add any further points that are at least distance 1 from all the points we have got, it can only be because every point of the sphere is *within* distance 1 of at least one of our chosen points. So the sphere has a 1-net consisting of only $4^n$ points, which is what we wanted to show.                                □

Lemmas 2.2 and 2.3 are routine calculations that can be done in many ways. We leave Lemma 2.3 to the dedicated reader. Lemma 2.2, which will be quoted throughout Lectures 8 and 9, is closely related to the Gaussian decay of the volume of the ball described in the last lecture. At least for smallish $\varepsilon$ (which is the interesting range) it can be proved as follows.

PROOF. The proportion of $S^{n-1}$ belonging to the cap $C(\varepsilon, u)$ equals the proportion of the solid ball that lies in the "spherical cone" illustrated in Figure 11.

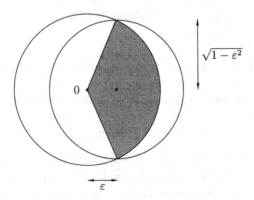

**Figure 11.** Estimating the area of a cap.

As is also illustrated, this spherical cone is contained in a ball of radius $\sqrt{1-\varepsilon^2}$ (if $\varepsilon \leq 1/\sqrt{2}$), so the ratio of its volume to that of the ball is at most

$$\left(1-\varepsilon^2\right)^{n/2} \leq e^{-n\varepsilon^2/2}. \qquad \square$$

In Lecture 8 we shall quote the upper estimate for areas of caps repeatedly. We shall in fact be using yet a third way to measure caps that differs very slightly from the $C(\varepsilon, u)$ description. The reader can easily check that the preceding argument yields the same estimate $e^{-n\varepsilon^2/2}$ for this other description.

## Lecture 3. Fritz John's Theorem

In the first lecture we saw that the cube and the cross-polytope lie at distance at most $\sqrt{n}$ from the Euclidean ball in $\mathbb{R}^n$, and that for the simplex, the distance is at most $n$. It is intuitively clear that these estimates cannot be improved. In this lecture we shall describe a strong sense in which this is as bad as things get. The theorem we shall describe was proved by Fritz John [1948].

John considered ellipsoids inside convex bodies. If $(e_j)_1^n$ is an orthonormal basis of $\mathbb{R}^n$ and $(\alpha_j)$ are positive numbers, the ellipsoid

$$\left\{ x : \sum_1^n \frac{\langle x, e_j \rangle^2}{\alpha_j^2} \leq 1 \right\}$$

has volume $v_n \prod \alpha_j$. John showed that each convex body contains a unique ellipsoid of largest volume and, more importantly, he *characterised* it. He showed that if $K$ is a symmetric convex body in $\mathbb{R}^n$ and $\mathcal{E}$ is its maximal ellipsoid then

$$K \subset \sqrt{n}\, \mathcal{E}.$$

Hence, after an affine transformation (one taking $\mathcal{E}$ to $B_2^n$) we can arrange that

$$B_2^n \subset K \subset \sqrt{n}B_2^n.$$

A nonsymmetric $K$ may require $nB_2^n$, like the simplex, rather than $\sqrt{n}B_2^n$.

John's characterisation of the maximal ellipsoid is best expressed after an affine transformation that takes the maximal ellipsoid to $B_2^n$. The theorem states that $B_2^n$ is the maximal ellipsoid in $K$ if a certain condition holds—roughly, that there be plenty of points of contact between the boundary of $K$ and that of the ball. See Figure 12.

THEOREM 3.1 (JOHN'S THEOREM). *Each convex body $K$ contains an unique ellipsoid of maximal volume. This ellipsoid is $B_2^n$ if and only if the following conditions are satisfied: $B_2^n \subset K$ and (for some $m$) there are Euclidean unit vectors $(u_i)_1^m$ on the boundary of $K$ and positive numbers $(c_i)_1^m$ satisfying*

$$\sum c_i u_i = 0 \qquad (3.1)$$

*and*

$$\sum c_i \langle x, u_i \rangle^2 = |x|^2 \quad \text{for each } x \in \mathbb{R}^n. \qquad (3.2)$$

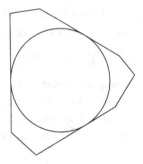

**Figure 12.** The maximal ellipsoid touches the boundary at many points.

According to the theorem the points at which the sphere touches $\partial K$ can be given a mass distribution whose centre of mass is the origin and whose inertia tensor is the identity matrix. Let's see where these conditions come from. The first condition, (3.1), guarantees that the $(u_i)$ do not all lie "on one side of the sphere". If they did, we could move the ball away from these contact points and blow it up a bit to obtain a larger ball in $K$. See Figure 13.

The second condition, (3.2), shows that the $(u_i)$ behave rather like an orthonormal basis in that we can resolve the Euclidean norm as a (weighted) sum of squares of inner products. Condition (3.2) is equivalent to the statement that

$$x = \sum c_i \langle x, u_i \rangle u_i \quad \text{for all } x.$$

This guarantees that the $(u_i)$ do not all lie close to a proper subspace of $\mathbb{R}^n$. If they did, we could shrink the ball a little in this subspace and expand it in an orthogonal direction, to obtain a larger ellipsoid inside $K$. See Figure 14.

Condition (3.2) is often written in matrix (or operator) notation as

$$\sum c_i \, u_i \otimes u_i = I_n \tag{3.3}$$

where $I_n$ is the identity map on $\mathbb{R}^n$ and, for any unit vector $u$, $u \otimes u$ is the rank-one orthogonal projection onto the span of $u$, that is, the map $x \mapsto \langle x, u \rangle u$. The trace of such an orthogonal projection is 1. By equating the traces of the

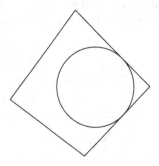

**Figure 13.** An ellipsoid where all contacts are on one side can grow.

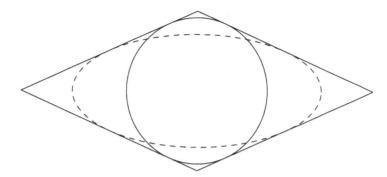

**Figure 14.** An ellipsoid (solid circle) whose contact points are all near one plane can grow.

matrices in the preceding equation, we obtain

$$\sum c_i = n.$$

In the case of a *symmetric* convex body, condition (3.1) is redundant, since we can take any sequence $(u_i)$ of contact points satisfying condition (3.2) and replace each $u_i$ by the pair $\pm u_i$ each with half the weight of the original.

Let's look at a few concrete examples. The simplest is the cube. For this body the maximal ellipsoid is $B_2^n$, as one would expect. The contact points are the standard basis vectors $(e_1, e_2, \ldots, e_n)$ of $\mathbb{R}^n$ and their negatives, and they satisfy

$$\sum_1^n e_i \otimes e_i = I_n.$$

That is, one can take all the weights $c_i$ equal to 1 in (3.2). See Figure 15.

The simplest nonsymmetric example is the simplex. In general, there is no natural way to place a regular simplex in $\mathbb{R}^n$, so there is no natural description of the contact points. Usually the best way to talk about an $n$-dimensional simplex is to realise it in $\mathbb{R}^{n+1}$: for example as the convex hull of the standard basis

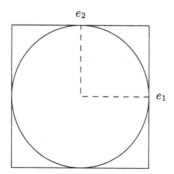

**Figure 15.** The maximal ellipsoid for the cube.

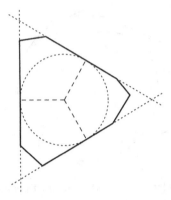

**Figure 16.** $K$ is contained in the convex body determined by the hyperplanes tangent to the maximal ellipsoid at the contact points.

vectors in $\mathbb{R}^{n+1}$. We leave it as an exercise for the reader to come up with a nice description of the contact points.

One may get a bit more of a feel for the second condition in John's Theorem by interpreting it as a rigidity condition. A sequence of unit vectors $(u_i)$ satisfying the condition (for some sequence $(c_i)$) has the property that if $T$ is a linear map of determinant 1, not all the images $Tu_i$ can have Euclidean norm less than 1.

John's characterisation immediately implies the inclusion mentioned earlier: if $K$ is symmetric and $\mathcal{E}$ is its maximal ellipsoid then $K \subset \sqrt{n}\,\mathcal{E}$. To check this we may assume $\mathcal{E} = B_2^n$. At each contact point $u_i$, the convex bodies $B_2^n$ and $K$ have the same supporting hyperplane. For $B_2^n$, the supporting hyperplane at any point $u$ is perpendicular to $u$. Thus if $x \in K$ we have $\langle x, u_i \rangle \leq 1$ for each $i$, and we conclude that $K$ is a subset of the convex body

$$C = \{x \in \mathbb{R}^n : \langle x, u_i \rangle \leq 1 \text{ for } 1 \leq i \leq m\}. \tag{3.4}$$

An example of this is shown in Figure 16.

In the symmetric case, the same argument shows that for each $x \in K$ we have $|\langle x, u_i \rangle| \leq 1$ for each $i$. Hence, for $x \in K$,

$$|x|^2 = \sum c_i \langle x, u_i \rangle^2 \leq \sum c_i = n.$$

So $|x| \leq \sqrt{n}$, which is exactly the statement $K \subset \sqrt{n}\,B_2^n$. We leave as a slightly trickier exercise the estimate $|x| \leq n$ in the nonsymmetric case.

PROOF OF JOHN'S THEOREM. The proof is in two parts, the harder of which is to show that if $B_2^n$ is *an* ellipsoid of largest volume, then we can find an appropriate system of weights on the contact points. The easier part is to show that if such a system of weights exists, then $B_2^n$ is the *unique* ellipsoid of maximal volume. We shall describe the proof only in the symmetric case, since the added complications in the general case add little to the ideas.

We begin with the easier part. Suppose there are unit vectors $(u_i)$ in $\partial K$ and numbers $(c_i)$ satisfying

$$\sum c_i\, u_i \otimes u_i = I_n.$$

Let

$$\mathcal{E} = \left\{ x : \sum_1^n \frac{\langle x, e_j \rangle^2}{\alpha_j^2} \leq 1 \right\}$$

be an ellipsoid in $K$, for some orthonormal basis $(e_j)$ and positive $\alpha_j$. We want to show that

$$\prod_1^n \alpha_j \leq 1$$

and that the product is equal to 1 only if $\alpha_j = 1$ for all $j$.

Since $\mathcal{E} \subset K$ we have that for each $i$ the hyperplane $\{x : \langle x, u_i \rangle = 1\}$ does not cut $\mathcal{E}$. This implies that each $u_i$ belongs to the *polar ellipsoid*

$$\left\{ y : \sum_1^n \alpha_j^2 \langle y, e_j \rangle^2 \leq 1 \right\}.$$

(The reader unfamiliar with duality is invited to check this.) So, for each $i$,

$$\sum_{j=1}^n \alpha_j^2 \langle u_i, e_j \rangle^2 \leq 1.$$

Hence

$$\sum_i c_i \sum_j \alpha_j^2 \langle u_i, e_j \rangle^2 \leq \sum c_i = n.$$

But the left side of the equality is just $\sum_j \alpha_j^2$, because, by condition (3.2), we have

$$\sum_i c_i \langle u_i, e_j \rangle^2 = |e_j|^2 = 1$$

for each $j$. Finally, the fact that the geometric mean does not exceed the arithmetic mean (the AM/GM inequality) implies that

$$\left( \prod \alpha_j^2 \right)^{1/n} \leq \frac{1}{n} \sum \alpha_j^2 \leq 1,$$

and there is equality in the first of these inequalities only if all $\alpha_j$ are equal to 1.

We now move to the harder direction. Suppose $B_2^n$ is an ellipsoid of largest volume in $K$. We want to show that there is a sequence of contact points $(u_i)$ and positive weights $(c_i)$ with

$$\frac{1}{n} I_n = \frac{1}{n} \sum c_i\, u_i \otimes u_i.$$

We already know that, if this is possible, we must have

$$\sum \frac{c_i}{n} = 1.$$

So our aim is to show that the matrix $I_n/n$ can be written as a convex combination of (a finite number of) matrices of the form $u \otimes u$, where each $u$ is a contact point. Since the space of matrices is finite-dimensional, the problem is simply to show that $I_n/n$ belongs to the convex hull of the set of all such rank-one matrices,

$$T = \{u \otimes u : u \text{ is a contact point}\}.$$

We shall aim to get a contradiction by showing that if $I_n/n$ is *not* in $T$, we can perturb the unit ball slightly to get a new ellipsoid in $K$ of larger volume than the unit ball.

Suppose that $I_n/n$ is not in $T$. Apply the separation theorem in the space of matrices to get a linear functional $\phi$ (on this space) with the property that

$$\phi\left(\frac{I_n}{n}\right) < \phi(u \otimes u) \tag{3.5}$$

for each contact point $u$. Observe that $\phi$ can be represented by an $n \times n$ matrix $H = (h_{jk})$, so that, for any matrix $A = (a_{jk})$,

$$\phi(A) = \sum_{jk} h_{jk} a_{jk}.$$

Since all the matrices $u \otimes u$ and $I_n/n$ are symmetric, we may assume the same for $H$. Moreover, since these matrices all have the same trace, namely 1, the inequality $\phi(I_n/n) < \phi(u \otimes u)$ will remain unchanged if we add a constant to each diagonal entry of $H$. So we may assume that the trace of $H$ is 0: but this says precisely that $\phi(I_n) = 0$.

Hence, unless the identity has the representation we want, we have found a symmetric matrix $H$ with zero trace for which

$$\sum_{jk} h_{jk}(u \otimes u)_{jk} > 0$$

for every contact point $u$. We shall use this $H$ to build a bigger ellipsoid inside $K$.

Now, for each vector $u$, the expression

$$\sum_{jk} h_{jk}(u \otimes u)_{jk}$$

is just the number $u^T H u$. For sufficiently small $\delta > 0$, the set

$$\mathcal{E}_\delta = \{x \in \mathbb{R}^n : x^T(I_n + \delta H)x \leq 1\}$$

is an ellipsoid and as $\delta$ tends to 0 these ellipsoids approach $B_2^n$. If $u$ is one of the original contact points, then

$$u^T(I_n + \delta H)u = 1 + \delta u^T H u > 1,$$

so $u$ does not belong to $\mathcal{E}_\delta$. Since the boundary of $K$ is compact (and the function $x \mapsto x^T H x$ is continuous) $\mathcal{E}_\delta$ will not contain any other point of $\partial K$ as long as

$\delta$ is sufficiently small. Thus, for such $\delta$, the ellipsoid $\mathcal{E}_\delta$ is strictly inside $K$ and some slightly expanded ellipsoid is inside $K$.

It remains to check that each $\mathcal{E}_\delta$ has volume at least that of $B_2^n$. If we denote by $(\mu_j)$ the eigenvalues of the symmetric matrix $I_n + \delta H$, the volume of $\mathcal{E}_\delta$ is $v_n / \prod \mu_j$, so the problem is to show that, for each $\delta$, we have $\prod \mu_j \leq 1$. What we know is that $\sum \mu_j$ is the trace of $I_n + \delta H$, which is $n$, since the trace of $H$ is 0. So the AM/GM inequality again gives

$$\prod \mu_j^{1/n} \leq \frac{1}{n} \sum \mu_j \leq 1,$$

as required.                                                                     $\square$

There is an analogue of John's Theorem that characterises the unique ellipsoid of minimal volume containing a given convex body. (The characterisation is almost identical, guaranteeing a resolution of the identity in terms of contact points of the body and the Euclidean sphere.) This minimal volume ellipsoid theorem can be deduced directly from John's Theorem by duality. It follows that, for example, the ellipsoid of minimal volume containing the cube $[-1, 1]^n$ is the obvious one: the ball of radius $\sqrt{n}$. It has been mentioned several times without proof that the distance of the cube from the Euclidean ball in $\mathbb{R}^n$ is exactly $\sqrt{n}$. We can now see this easily: the ellipsoid of minimal volume outside the cube has volume $(\sqrt{n})^n$ times that of the ellipsoid of maximal volume inside the cube. So we cannot sandwich the cube between homothetic ellipsoids unless the outer one is at least $\sqrt{n}$ times the inner one.

We shall be using John's Theorem several times in the remaining lectures. At this point it is worth mentioning important extensions of the result. We can view John's Theorem as a description of those linear maps from Euclidean space to a normed space (whose unit ball is $K$) that have largest determinant, subject to the condition that they have norm at most 1: that is, that they map the Euclidean ball into $K$. There are many other norms that can be put on the space of linear maps. John's Theorem is the starting point for a general theory that builds ellipsoids related to convex bodies by maximising determinants subject to other constraints on linear maps. This theory played a crucial role in the development of convex geometry over the last 15 years. This development is described in detail in [Tomczak-Jaegermann 1988].

## Lecture 4. Volume Ratios and Spherical Sections of the Octahedron

In the second lecture we saw that the $n$-dimensional cube has almost spherical sections of dimension about $\log n$ but not more. In this lecture we will examine the corresponding question for the $n$-dimensional cross-polytope $B_1^n$. In itself, this body is as far from the Euclidean ball as is the cube in $\mathbb{R}^n$: its distance from the ball, in the sense described in Lecture 2 is $\sqrt{n}$. Remarkably, however, it has almost spherical sections whose dimension is about $\frac{1}{2}n$. We shall deduce this from what is perhaps an even more surprising statement concerning intersections of bodies. Recall that $B_1^n$ contains the Euclidean ball of radius $\frac{1}{\sqrt{n}}$. If $U$ is an orthogonal transformation of $\mathbb{R}^n$ then $UB_1^n$ also contains this ball and hence so does the intersection $B_1^n \cap UB_1^n$. But, whereas $B_1^n$ does not lie in any Euclidean ball of radius less than 1, we have the following theorem [Kašin 1977]:

THEOREM 4.1. *For each $n$, there is an orthogonal transformation $U$ for which the intersection $B_1^n \cap UB_1^n$ is contained in the Euclidean ball of radius $32/\sqrt{n}$ (and contains the ball of radius $1/\sqrt{n}$).*

(The constant 32 can easily be improved: the important point is that it is independent of the dimension $n$.) The theorem states that the intersection of just two copies of the $n$-dimensional octahedron may be approximately spherical. Notice that if we tried to approximate the Euclidean ball by intersecting rotated copies of the cube, we would need exponentially many in the dimension, because the cube has only $2n$ facets and our approximation needs exponentially many facets. In contrast, the octahedron has a much larger number of facets, $2^n$: but of course we need to do a lot more than just count facets in order to prove Theorem 4.1. Before going any further we should perhaps remark that the cube has a property that corresponds to Theorem 4.1. If $Q$ is the cube and $U$ is the same orthogonal transformation as in the theorem, the convex hull

$$\text{conv}(Q \cup UQ)$$

is at distance at most 32 from the Euclidean ball.

In spirit, the argument we shall use to establish Theorem 4.1 is Kašin's original one but, following [Szarek 1978], we isolate the main ingredient and we organise the proof along the lines of [Pisier 1989]. Some motivation may be helpful. The ellipsoid of maximal volume inside $B_1^n$ is the Euclidean ball of radius $\frac{1}{\sqrt{n}}$. (See Figure 3.) There are $2^n$ points of contact between this ball and the boundary of $B_1^n$: namely, the points of the form

$$\left( \pm\frac{1}{n}, \pm\frac{1}{n}, \ldots, \pm\frac{1}{n} \right),$$

one in the middle of each facet of $B_1^n$. The vertices,

$$(\pm 1, 0, 0, \ldots, 0), \ldots, (0, 0, \ldots, 0, \pm 1),$$

are the points of $B_1^n$ furthest from the origin. We are looking for a rotation $UB_1^n$ whose facets chop off the spikes of $B_1^n$ (or vice versa). So we want to know that the points of $B_1^n$ at distance about $1/\sqrt{n}$ from the origin are fairly typical, while those at distance 1 are atypical.

For a unit vector $\theta \in S^{n-1}$, let $r(\theta)$ be the radius of $B_1^n$ in the direction $\theta$,

$$r(\theta) = \frac{1}{\|\theta\|_1} = \left( \sum_1^n |\theta_i| \right)^{-1}.$$

In the first lecture it was explained that the volume of $B_1^n$ can be written

$$v_n \int_{S^{n-1}} r(\theta)^n d\sigma$$

and that it is equal to $2^n/n!$. Hence

$$\int_{S^{n-1}} r(\theta)^n d\sigma = \frac{2^n}{n! \, v_n} \le \left( \frac{2}{\sqrt{n}} \right)^n.$$

Since the average of $r(\theta)^n$ is at most $\left( 2/\sqrt{n} \right)^n$, the value of $r(\theta)$ cannot often be much more than $2/\sqrt{n}$. This feature of $B_1^n$ is captured in the following definition of Szarek.

DEFINITION. Let $K$ be a convex body in $\mathbb{R}^n$. The *volume ratio* of $K$ is

$$\mathrm{vr}(K) = \left( \frac{\mathrm{vol}(K)}{\mathrm{vol}(\mathcal{E})} \right)^{1/n},$$

where $\mathcal{E}$ is the ellipsoid of maximal volume in $K$.

The preceding discussion shows that $\mathrm{vr}(B_1^n) \le 2$ for all $n$. Contrast this with the cube in $\mathbb{R}^n$, whose volume ratio is about $\sqrt{n}/2$. The only property of $B_1^n$ that we shall use to prove Kašin's Theorem is that its volume ratio is at most 2. For convenience, we scale everything up by a factor of $\sqrt{n}$ and prove the following.

THEOREM 4.2.   *Let $K$ be a symmetric convex body in $\mathbb{R}^n$ that contains the Euclidean unit ball $B_2^n$ and for which*

$$\left( \frac{\mathrm{vol}(K)}{\mathrm{vol}(B_2^n)} \right)^{1/n} = R.$$

*Then there is an orthogonal transformation $U$ of $\mathbb{R}^n$ for which*

$$K \cap UK \subset 8R^2 B_2^n.$$

PROOF. It is convenient to work with the norm on $\mathbb{R}^n$ whose unit ball is $K$. Let $\| \cdot \|$ denote this norm and $| \cdot |$ the standard Euclidean norm. Notice that, since $B_2^n \subset K$, we have $\|x\| \le |x|$ for all $x \in \mathbb{R}^n$.

The radius of the body $K \cap UK$ in a given direction is the minimum of the radii of $K$ and $UK$ in that direction. So the norm corresponding to the body $K \cap UK$ is the *maximum* of the norms corresponding to $K$ and $UK$. We need to find an orthogonal transformation $U$ with the property that

$$\max \left( \|U\theta\|, \|\theta\| \right) \ge \frac{1}{8R^2}$$

for every $\theta \in S^{n-1}$. Since the maximum of two numbers is at least their average, it will suffice to find $U$ with

$$\frac{\|U\theta\| + \|\theta\|}{2} \ge \frac{1}{8R^2} \quad \text{for all } \theta.$$

For each $x \in \mathbb{R}^n$ write $N(x)$ for the average $\frac{1}{2}(\|Ux\| + \|x\|)$. One sees immediately that $N$ is a norm (that is, it satisfies the triangle inequality) and that $N(x) \leq |x|$ for every $x$, since $U$ preserves the Euclidean norm. We shall show in a moment that there is a $U$ for which

$$\int_{S^{n-1}} \frac{1}{N(\theta)^{2n}} \, d\sigma \leq R^{2n}. \tag{4.1}$$

This says that $N(\theta)$ is large on average: we want to deduce that it is large everywhere.

Let $\phi$ be a point of the sphere and write $N(\phi) = t$, for $0 < t \leq 1$. The crucial point will be that, if $\theta$ is close to $\phi$, then $N(\theta)$ cannot be much more than $t$. To be precise, if $|\theta - \phi| \leq t$ then

$$N(\theta) \leq N(\phi) + N(\theta - \phi) \leq t + |\theta - \phi| \leq 2t.$$

Hence, $N(\theta)$ is at most $2t$ for every $\theta$ in a spherical cap of radius $t$ about $\phi$. From Lemma 2.3 we know that this spherical cap has measure at least

$$\frac{1}{2}\left(\frac{t}{2}\right)^{n-1} \geq \left(\frac{t}{2}\right)^n.$$

So $1/N(\theta)^{2n}$ is at least $1/(2t)^{2n}$ on a set of measure at least $(t/2)^n$. Therefore

$$\int_{S^{n-1}} \frac{1}{N(\theta)^{2n}} \, d\sigma \geq \frac{1}{(2t)^{2n}}\left(\frac{t}{2}\right)^n = \frac{1}{2^{3n}t^n}.$$

By (4.1), the integral is at most $R^{2n}$, so $t \geq 1/(8R^2)$. Thus our arbitrary point $\phi$ satisfies

$$N(\phi) \geq \frac{1}{8R^2}.$$

It remains to find $U$ satisfying (4.1). Now, for any $\theta$, we have

$$N(\theta)^2 = \left(\frac{\|U\theta\| + \|\theta\|}{2}\right)^2 \geq \|U\theta\| \, \|\theta\|,$$

so it will suffice to find a $U$ for which

$$\int_{S^{n-1}} \frac{1}{\|U\theta\|^n \, \|\theta\|^n} \, d\sigma \leq R^{2n}. \tag{4.2}$$

The hypothesis on the volume of $K$ can be written in terms of the norm as

$$\int_{S^{n-1}} \frac{1}{\|\theta\|^n} \, d\sigma = R^n.$$

The group of orthogonal transformations carries an invariant probability measure. This means that we can average a function over the group in a natural way. In particular, if $f$ is a function on the sphere and $\theta$ is some point on the sphere, the average over orthogonal $U$ of the value $f(U\theta)$ is just the average of $f$ on the sphere: averaging over $U$ mimics averaging over the sphere:

$$\text{ave}_U \, f(U\theta) = \int_{S^{n-1}} f(\phi) \, d\sigma(\phi).$$

Hence,

$$\text{ave}_U \int_{S^{n-1}} \frac{1}{\|U\theta\|^n . \|\theta\|^n} \, d\sigma(\theta) = \int_{S^{n-1}} \left( \text{ave}_U \frac{1}{\|U\theta\|^n} \right) \frac{1}{\|\theta\|^n} \, d\sigma(\theta)$$

$$= \int_{S^{n-1}} \left( \int_{S^{n-1}} \frac{1}{\|\phi\|^n} \, d\sigma(\phi) \right) \frac{1}{\|\theta\|^n} \, d\sigma(\theta)$$

$$= \left( \int_{S^{n-1}} \frac{1}{\|\theta\|^n} \, d\sigma(\theta) \right)^2 = R^{2n}.$$

Since the average over all $U$ of the integral is at most $R^{2n}$, there is at least one $U$ for which the integral is at most $R^{2n}$. This is exactly inequality (4.2). $\quad\square$

The choice of $U$ in the preceding proof is a random one. The proof does not in any way tell us how to find an explicit $U$ for which the integral is small. In the case of a general body $K$, this is hardly surprising, since we are assuming nothing about how the volume of $K$ is distributed. But, in view of the earlier remarks about facets of $B_1^n$ chopping off spikes of $UB_1^n$, it is tempting to think that for the particular body $B_1^n$ we might be able to write down an appropriate $U$ explicitly. In two dimensions the best choice of $U$ is obvious: we rotate the diamond through $45°$ and after intersection we have a regular octagon as shown in Figure 17.

The most natural way to try to generalise this to higher dimensions is to look for a $U$ such that each vertex of $UB_1^n$ is exactly aligned through the centre of a facet of $B_1^n$: that is, for each standard basis vector $e_i$ of $\mathbb{R}^n$, $Ue_i$ is a multiple of one of the vectors $(\pm\frac{1}{n}, \ldots, \pm\frac{1}{n})$. (The multiple is $\sqrt{n}$ since $Ue_i$ has length 1.) Thus we are looking for an $n \times n$ orthogonal matrix $U$ each of whose entries is

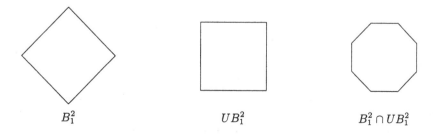

$$B_1^2 \qquad\qquad UB_1^2 \qquad\qquad B_1^2 \cap UB_1^2$$

**Figure 17.** The best choice for $U$ in two dimensions is a $45°$ rotation.

$\pm 1/\sqrt{n}$. Such matrices, apart from the factor $\sqrt{n}$, are called *Hadamard matrices*. In what dimensions do they exist? In dimensions 1 and 2 there are the obvious

$$(1) \quad \text{and} \quad \begin{pmatrix} \frac{1}{\sqrt{2}} & \frac{1}{\sqrt{2}} \\ \frac{1}{\sqrt{2}} & -\frac{1}{\sqrt{2}} \end{pmatrix}.$$

It is not too hard to show that in larger dimensions a Hadamard matrix cannot exist unless the dimension is a multiple of 4. It is an open problem to determine whether they exist in all dimensions that are multiples of 4. They are known to exist, for example, if the dimension is a power of 2: these examples are known as the Walsh matrices.

In spite of this, it seems extremely unlikely that one might prove Kašin's Theorem using Hadamard matrices. The Walsh matrices certainly do not give anything smaller than $n^{-1/4}$; pretty miserable compared with $n^{-1/2}$. There are some good reasons, related to Ramsey theory, for believing that one cannot expect to find genuinely explicit matrices of any kind that would give the right estimates.

Let's return to the question with which we opened the lecture and see how Theorem 4.1 yields almost spherical sections of octahedra. We shall show that, for each $n$, the $2n$-dimensional octahedron has an $n$-dimensional slice which is within distance 32 of the ($n$-dimensional) Euclidean ball. By applying the argument of the theorem to $B_1^n$, we obtain an $n \times n$ orthogonal matrix $U$ such that

$$\|Ux\|_1 + \|x\|_1 \geq \frac{\sqrt{n}}{16}|x|$$

for every $x \in \mathbb{R}^n$, where $\|\cdot\|_1$ denotes the $\ell_1$ norm. Now consider the map $T : \mathbb{R}^n \to \mathbb{R}^{2n}$ with matrix $\binom{U}{I}$. For each $x \in \mathbb{R}^n$, the norm of $Tx$ in $\ell_1^{2n}$ is

$$\|Tx\|_1 = \|Ux\|_1 + \|x\|_1 \geq \frac{\sqrt{n}}{16}|x|.$$

On the other hand, the Euclidean norm of $Tx$ is

$$|Tx| = \sqrt{|Ux|^2 + |x|^2} = \sqrt{2}|x|.$$

So, if $y$ belongs to the image $T\mathbb{R}^n$, then, setting $y = Tx$,

$$\|y\|_1 \geq \frac{\sqrt{n}}{16}|x| = \frac{\sqrt{n}}{16\sqrt{2}}|y| = \frac{\sqrt{2n}}{32}|y|.$$

By the Cauchy–Schwarz inequality, we have $\|y\|_1 \leq \sqrt{2n}|y|$, so the slice of $B_1^{2n}$ by the subspace $T\mathbb{R}^n$ has distance at most 32 from $B_2^n$, as we wished to show.

A good deal of work has been done on embedding of other subspaces of $L_1$ into $\ell_1$-spaces of low dimension, and more generally subspaces of $L_p$ into low-dimensional $\ell_p$, for $1 < p < 2$. The techniques used come from probability theory: $p$-stable random variables, bootstrapping of probabilities and deviation estimates. We shall be looking at applications of the latter in Lectures 8 and 9.

The major references are [Johnson and Schechtman 1982; Bourgain et al. 1989; Talagrand 1990].

Volume ratios have been studied in considerable depth. They have been found to be closely related to the so-called cotype-2 property of normed spaces: this relationship is dealt with comprehensively in [Pisier 1989]. In particular, Bourgain and Milman [1987] showed that a bound for the cotype-2 constant of a space implies a bound for the volume ratio of its unit ball. This demonstrated, among other things, that there is a uniform bound for the volume ratios of slices of octahedra of all dimensions. A sharp version of this result was proved in [Ball 1991]: namely, that for each $n$, $B_1^n$ has largest volume ratio among the balls of $n$-dimensional subspaces of $L_1$. The proof uses techniques that will be discussed in Lecture 6.

This is a good point to mention a result of Milman [1985] that looks superficially like the results of this lecture but lies a good deal deeper. We remarked that while we can almost get a sphere by intersecting two copies of $B_1^n$, this is very far from possible with two cubes. Conversely, we can get an almost spherical convex hull of two cubes but not of two copies of $B_1^n$. The *QS-Theorem* (an abbreviation for "quotient of a subspace") states that if we combine the two operations, intersection and convex hull, we can get a sphere no matter what body we start with.

THEOREM 4.3 (QS-THEOREM). *There is a constant $M$ (independent of everything) such that, for any symmetric convex body $K$ of any dimension, there are linear maps $Q$ and $S$ and an ellipsoid $\mathcal{E}$ with the following property: if $\tilde{K} = \mathrm{conv}(K \cup QK)$, then*

$$\mathcal{E} \subset \tilde{K} \cap S\tilde{K} \subset M\mathcal{E}.$$

# Lecture 5. The Brunn–Minkowski Inequality and Its Extensions

In this lecture we shall introduce one of the most fundamental principles in convex geometry: the Brunn–Minkowski inequality. In order to motivate it, let's begin with a simple observation concerning convex sets in the plane. Let $K \subset \mathbb{R}^2$ be such a set and consider its slices by a family of parallel lines, for example those parallel to the $y$-axis. If the line $x = r$ meets $K$, call the length of the slice $v(r)$.

The graph of $v$ is obtained by shaking $K$ down onto the $x$-axis like a deck of cards (of different lengths). This is shown in Figure 18. It is easy to see that the function $v$ is concave on its support. Towards the end of the last century, Brunn investigated what happens if we try something similar in higher dimensions.

Figure 19 shows an example in three dimensions. The central, hexagonal, slice has larger volume than the triangular slices at the ends: each triangular slice can be decomposed into four smaller triangles, while the hexagon is a union of six such triangles. So our first guess might be that the slice area is a concave

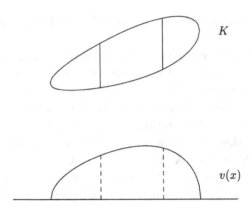

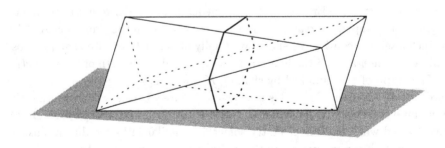

**Figure 18.** Shaking down a convex body.

function, just as slice length was concave for sets in the plane. That this is not always so can be seen by considering slices of a cone, parallel to its base: see Figure 20.

Since the area of a slice varies as the square of its distance from the cone's vertex, the area function obtained looks like a piece of the curve $y = x^2$, which is certainly not concave. However, it is reasonable to guess that the cone is an extremal example, since it is "only just" a convex body: its curved surface is "made up of straight lines". For this body, the square root of the slice function just manages to be concave on its support (since its graph is a line segment). So our second guess might be that for a convex body in $\mathbb{R}^3$, a slice-area function has a *square-root* that is concave on its support. This was proved by Brunn using an elegant symmetrisation method. His argument works just as well in higher dimensions to give the following result for the $(n-1)$-dimensional volumes of slices of a body in $\mathbb{R}^n$.

**Figure 19.** A polyhedron in three dimensions. The faces at the right and left are parallel.

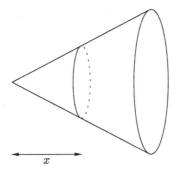

**Figure 20.** The area of a cone's section increases with $x^2$.

THEOREM 5.1 (BRUNN). *Let $K$ be a convex body in $\mathbb{R}^n$, let $u$ be a unit vector in $\mathbb{R}^n$, and for each $r$ let $H_r$ be the hyperplane*

$$\{x \in \mathbb{R}^n : \langle x, u \rangle = r\}.$$

*Then the function*

$$r \mapsto \operatorname{vol}(K \cap H_r)^{1/(n-1)}$$

*is concave on its support.*

One consequence of this is that if $K$ is centrally symmetric, the largest slice perpendicular to a given direction is the central slice (since an even concave function is largest at 0). This is the situation in Figure 19.

Brunn's Theorem was turned from an elegant curiosity into a powerful tool by Minkowski. His reformulation works in the following way. Consider three parallel slices of a convex body in $\mathbb{R}^n$ at positions $r$, $s$ and $t$, where $s = (1 - \lambda)r + \lambda t$ for some $\lambda \in (0, 1)$. This is shown in Figure 21.

Call the slices $A_r$, $A_s$, and $A_t$ and think of them as subsets of $\mathbb{R}^{n-1}$. If $x \in A_r$ and $y \in A_t$, the point $(1 - \lambda)x + \lambda y$ belongs to $A_s$: to see this, join the points $(r, x)$ and $(t, y)$ in $\mathbb{R}^n$ and observe that the resulting line segment crosses $A_s$ at $(s, (1 - \lambda)x + \lambda y)$. So $A_s$ includes a new set

$$(1 - \lambda)A_r + \lambda A_t := \{(1 - \lambda)x + \lambda y : x \in A_r, y \in A_t\}.$$

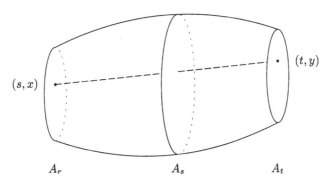

**Figure 21.** The section $A_s$ contains the weighted average of $A_r$ and $A_t$.

(This way of using the addition in $\mathbb{R}^n$ to define an addition of sets is called *Minkowski addition.*) Brunn's Theorem says that the volumes of the three sets $A_r$, $A_s$, and $A_t$ in $\mathbb{R}^{n-1}$ satisfy

$$\operatorname{vol}(A_s)^{1/(n-1)} \geq (1-\lambda)\operatorname{vol}(A_r)^{1/(n-1)} + \lambda\operatorname{vol}(A_t)^{1/(n-1)}.$$

The Brunn–Minkowski inequality makes explicit the fact that all we really know about $A_s$ is that it includes the Minkowski combination of $A_r$ and $A_t$. Since we have now eliminated the role of the ambient space $\mathbb{R}^n$, it is natural to rewrite the inequality with $n$ in place of $n-1$.

THEOREM 5.2 (BRUNN–MINKOWSKI INEQUALITY). *If $A$ and $B$ are nonempty compact subsets of $\mathbb{R}^n$ then*

$$\operatorname{vol}((1-\lambda)A + \lambda B)^{1/n} \geq (1-\lambda)\operatorname{vol}(A)^{1/n} + \lambda\operatorname{vol}(B)^{1/n}.$$

(The hypothesis that $A$ and $B$ be nonempty corresponds in Brunn's Theorem to the restriction of a function to its support.) It should be remarked that the inequality is stated for general compact sets, whereas the early proofs gave the result only for convex sets. The first complete proof for the general case seems to be in [Lîusternik 1935].

To get a feel for the advantages of Minkowski's formulation, let's see how it implies the classical isoperimetric inequality in $\mathbb{R}^n$.

THEOREM 5.3 (ISOPERIMETRIC INEQUALITY). *Among bodies of a given volume, Euclidean balls have least surface area.*

PROOF. Let $C$ be a compact set in $\mathbb{R}^n$ whose volume is equal to that of $B_2^n$, the Euclidean ball of radius 1. The surface "area" of $C$ can be written

$$\operatorname{vol}(\partial C) = \lim_{\varepsilon \to 0} \frac{\operatorname{vol}(C + \varepsilon B_2^n) - \operatorname{vol}(C)}{\varepsilon},$$

as shown in Figure 22. By the Brunn–Minkowski inequality,

$$\operatorname{vol}(C + \varepsilon B_2^n)^{1/n} \geq \operatorname{vol}(C)^{1/n} + \varepsilon \operatorname{vol}(B_2^n)^{1/n}.$$

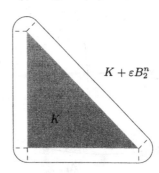

**Figure 22.** Expressing the area as a limit of volume increments.

Hence
$$\mathrm{vol}\,(C + \varepsilon B_2^n) \geq \left(\mathrm{vol}\,(C)^{1/n} + \varepsilon\,\mathrm{vol}\,(B_2^n)^{1/n}\right)^n$$
$$\geq \mathrm{vol}\,(C) + n\varepsilon\,\mathrm{vol}\,(C)^{(n-1)/n}\,\mathrm{vol}\,(B_2^n)^{1/n}.$$

So
$$\mathrm{vol}(\partial C) \geq n\,\mathrm{vol}(C)^{(n-1)/n}\,\mathrm{vol}(B_2^n)^{1/n}.$$

Since $C$ and $B_2^n$ have the same volume, this shows that $\mathrm{vol}(\partial C) \geq n\,\mathrm{vol}(B_2^n)$, and the latter equals $\mathrm{vol}(\partial B_2^n)$, as we saw in Lecture 1.    □

This relationship between the Brunn–Minkowski inequality and the isoperimetric inequality will be explored in a more general context in Lecture 8.

The Brunn–Minkowski inequality has an alternative version that is formally weaker. The AM/GM inequality shows that, for $\lambda$ in $(0,1)$,
$$(1 - \lambda)\,\mathrm{vol}(A)^{1/n} + \lambda\,\mathrm{vol}(B)^{1/n} \geq \mathrm{vol}(A)^{(1-\lambda)/n}\,\mathrm{vol}(B)^{\lambda/n}.$$

So the Brunn–Minkowski inequality implies that, for compact sets $A$ and $B$ and $\lambda \in (0,1)$,
$$\mathrm{vol}((1 - \lambda)A + \lambda B) \geq \mathrm{vol}(A)^{1-\lambda}\,\mathrm{vol}(B)^{\lambda}. \tag{5.1}$$

Although this multiplicative Brunn–Minkowski inequality is weaker than the Brunn–Minkowski inequality for *particular* $A$, $B$, and $\lambda$, if one knows (5.1) for *all* $A$, $B$, and $\lambda$ one can easily deduce the Brunn–Minkowski inequality for all $A$, $B$, and $\lambda$. This deduction will be left for the reader.

Inequality (5.1) has certain advantages over the Brunn–Minkowski inequality.

(i) We no longer need to stipulate that $A$ and $B$ be nonempty, which makes the inequality easier to use.

(ii) The dimension $n$ has disappeared.

(iii) As we shall see, the multiplicative inequality lends itself to a particularly simple proof because it has a generalisation from sets to functions.

Before we describe the functional Brunn–Minkowski inequality let's just remark that the multiplicative Brunn–Minkowski inequality can be reinterpreted back in the setting of Brunn's Theorem: if $r \mapsto v(r)$ is a function obtained by scanning a convex body with parallel hyperplanes, then $\log v$ is a concave function (with the usual convention regarding $-\infty$).

In order to move toward a functional generalisation of the multiplicative Brunn–Minkowski inequality let's reinterpret inequality (5.1) in terms of the characteristic functions of the sets involved. Let $f$, $g$, and $m$ denote the characteristic functions of $A$, $B$, and $(1 - \lambda)A + \lambda B$ respectively; so, for example, $f(x) = 1$ if $x \in A$ and $0$ otherwise. The volumes of $A$, $B$, and $(1 - \lambda)A + \lambda B$ are the integrals $\int_{\mathbb{R}^n} f$, $\int_{\mathbb{R}^n} g$, and $\int_{\mathbb{R}^n} m$. The Brunn–Minkowski inequality says that
$$\int m \geq \left(\int f\right)^{1-\lambda}\left(\int g\right)^{\lambda}.$$

But what is the relationship between $f$, $g$, and $m$ that guarantees its truth? If $f(x) = 1$ and $g(y) = 1$ then $x \in A$ and $y \in B$, so

$$(1 - \lambda)x + \lambda y \in (1 - \lambda)A + \lambda B,$$

and hence $m\left((1 - \lambda)x + \lambda y\right) = 1$. This certainly ensures that

$$m\left((1 - \lambda)x + \lambda y\right) \geq f(x)^{1-\lambda}g(y)^{\lambda} \quad \text{for any } x \text{ and } y \text{ in } \mathbb{R}^n.$$

This inequality for the three functions at least has a homogeneity that matches the desired inequality for the integrals. In a series of papers, Prékopa and Leindler proved that this homogeneity is enough.

THEOREM 5.4 (THE PRÉKOPA–LEINDLER INEQUALITY). *If $f$, $g$ and $m$ are nonnegative measurable functions on $\mathbb{R}^n$, $\lambda \in (0,1)$ and for all $x$ and $y$ in $\mathbb{R}^n$,*

$$m\left((1 - \lambda)x + \lambda y\right) \geq f(x)^{1-\lambda}g(y)^{\lambda} \tag{5.2}$$

*then*

$$\int m \geq \left(\int f\right)^{1-\lambda}\left(\int g\right)^{\lambda}.$$

It is perhaps helpful to notice that the Prékopa–Leindler inequality looks like Hölder's inequality, backwards. If $f$ and $g$ were given and we set

$$m(z) = f(z)^{1-\lambda}g(z)^{\lambda}$$

(for each $z$), then Hölder's inequality says that

$$\int m \leq \left(\int f\right)^{1-\lambda}\left(\int g\right)^{\lambda}.$$

(Hölder's inequality is often written with $1/p$ instead of $1 - \lambda$, $1/q$ instead of $\lambda$, and $f$, $g$ replaced by $F^p$, $G^q$.) The difference between Prékopa–Leindler and Hölder is that, in the former, the value $m(z)$ may be much larger since it is a supremum over many pairs $(x, y)$ satisfying $z = (1 - \lambda)x + \lambda y$ rather than just the pair $(z, z)$.

Though it generalises the Brunn–Minkowski inequality, the Prékopa–Leindler inequality is a good deal simpler to prove, once properly formulated. The argument we shall use seems to have appeared first in [Brascamp and Lieb 1976b]. The crucial point is that the passage from sets to functions allows us to prove the inequality by induction on the dimension, using only the one-dimensional case. We pay the small price of having to do a bit extra for this case.

PROOF OF THE PRÉKOPA–LEINDLER INEQUALITY. We start by checking the one-dimensional Brunn–Minkowski inequality. Suppose $A$ and $B$ are nonempty measurable subsets of the line. Using $|\cdot|$ to denote length, we want to show that

$$|(1 - \lambda)A + \lambda B| \geq (1 - \lambda)|A| + \lambda|B|.$$

We may assume that $A$ and $B$ are compact and we may shift them so that the right-hand end of $A$ and the left-hand end of $B$ are both at 0. The set $(1 - \lambda)A + \lambda B$ now includes the essentially disjoint sets $(1 - \lambda)A$ and $\lambda B$, so its length is at least the sum of the lengths of these sets.

Now suppose we have nonnegative integrable functions $f$, $g$, and $m$ on the line, satisfying condition (5.2). We may assume that $f$ and $g$ are bounded. Since the inequality to be proved has the same homogeneity as the hypothesis (5.2), we may also assume that $f$ and $g$ are normalised so that $\sup f = \sup g = 1$. By Fubini's Theorem, we can write the integrals of $f$ and $g$ as integrals of the lengths of their level sets:

$$\int f(x)\,dx = \int_0^1 |(f \geq t)|\,dt,$$

and similarly for $g$. If $f(x) \geq t$ and $g(y) \geq t$ then $m\left((1 - \lambda)x + \lambda y\right) \geq t$. So we have the inclusion

$$(m \geq t) \supset (1 - \lambda)(f \geq t) + \lambda(g \geq t).$$

For $0 \leq t < 1$ the sets on the right are nonempty so the one-dimensional Brunn–Minkowski inequality shows that

$$|(m \geq t)| \geq (1 - \lambda)\,|(f \geq t)| + \lambda\,|(g \geq t)|\,.$$

Integrating this inequality from 0 to 1 we get

$$\int m \geq (1 - \lambda) \int f + \lambda \int g,$$

and the latter expression is at least

$$\left(\int f\right)^{1-\lambda} \left(\int g\right)^{\lambda}$$

by the AM/GM inequality. This does the one-dimensional case.

The induction that takes us into higher dimensions is quite straightforward, so we shall just sketch the argument for sets in $\mathbb{R}^n$, rather than functions. Suppose $A$ and $B$ are two such sets and, for convenience, write

$$C = (1 - \lambda)A + \lambda B.$$

Choose a unit vector $u$ and, as before, let $H_r$ be the hyperplane

$$\{x \in \mathbb{R}^n : \langle x, u \rangle = r\}$$

perpendicular to $u$ at "position" $r$. Let $A_r$ denote the slice $A \cap H_r$ and similarly for $B$ and $C$, and regard these as subsets of $\mathbb{R}^{n-1}$. If $r$ and $t$ are real numbers, and if $s = (1-\lambda)r + \lambda t$, the slice $C_s$ includes $(1-\lambda)A_r + \lambda B_t$. (This is reminiscent

of the earlier argument relating Brunn's Theorem to Minkowski's reformulation.)
By the inductive hypothesis in $\mathbb{R}^{n-1}$,

$$\operatorname{vol}(C_s) \geq \operatorname{vol}(A_r)^{1-\lambda} \cdot \operatorname{vol}(B_t)^{\lambda}.$$

Let $f$, $g$, and $m$ be the functions on the line, given by

$$f(x) = \operatorname{vol}(A_x), \quad g(x) = \operatorname{vol}(B_x), \quad m(x) = \operatorname{vol}(C_x).$$

Then, for $r$, $s$, and $t$ as above,

$$m(s) \geq f(r)^{1-\lambda} g(t)^{\lambda}.$$

By the one-dimensional Prékopa–Leindler inequality,

$$\int m \geq \left( \int f \right)^{1-\lambda} \left( \int g \right)^{\lambda}.$$

But this is exactly the statement $\operatorname{vol}(C) \geq \operatorname{vol}(A)^{1-\lambda} \operatorname{vol}(B)^{\lambda}$, so the inductive
step is complete.                                                                        □

The proof illustrates clearly why the Prékopa–Leindler inequality makes things
go smoothly. Although we only carried out the induction for *sets*, we required
the one-dimensional result for the *functions* we get by scanning sets in $\mathbb{R}^n$.

   To close this lecture we remark that the Brunn–Minkowski inequality has numerous extensions and variations, not only in convex geometry, but in combinatorics and information theory as well. One of the most surprising and delightful
is a theorem of Busemann [1949].

THEOREM 5.5 (BUSEMANN). *Let $K$ be a symmetric convex body in $\mathbb{R}^n$, and
for each unit vector $u$ let $r(u)$ be the volume of the slice of $K$ by the subspace
orthogonal to $u$. Then the body whose radius in each direction $u$ is $r(u)$ is itself
convex.*

The Brunn–Minkowski inequality is the starting point for a highly developed
classical theory of convex geometry. We shall barely touch upon the theory in
these notes. A comprehensive reference is the recent book [Schneider 1993].

## Lecture 6. Convolutions and Volume Ratios:
## The Reverse Isoperimetric Problem

   In the last lecture we saw how to deduce the classical isoperimetric inequality
in $\mathbb{R}^n$ from the Brunn–Minkowski inequality. In this lecture we will answer the
reverse question. This has to be phrased a bit carefully, since there is no upper
limit to the surface area of a body of given volume, even if we restrict attention
to convex bodies. (Consider a very thin pancake.) For this reason it is natural to
consider affine equivalence classes of convex bodies, and the question becomes:
given a convex body, how small can we make its surface area by applying an

affine (or linear) transformation that preserves volume? The answer is provided by the following theorem from [Ball 1991].

THEOREM 6.1. *Let $K$ be a convex body and $T$ a regular solid simplex in $\mathbb{R}^n$. Then there is an affine image of $K$ whose volume is the same as that of $T$ and whose surface area is no larger than that of $T$.*

Thus, modulo affine transformations, simplices have the largest surface area among convex bodies of a given volume. If $K$ is assumed to be centrally symmetric then the estimate can be strengthened: the cube is extremal among symmetric bodies. A detailed proof of Theorem 6.1 would be too long for these notes. We shall instead describe how the symmetric case is proved, since this is considerably easier but illustrates the most important ideas.

Theorem 6.1 and the symmetric analogue are both deduced from volume-ratio estimates. In the latter case the statement is that among symmetric convex bodies, the cube has largest volume ratio. Let's see why this solves the reverse isoperimetric problem. If $Q$ is any cube, the surface area and volume of $Q$ are related by

$$\text{vol}(\partial Q) = 2n \, \text{vol}(Q)^{(n-1)/n}.$$

We wish to show that any other convex body $K$ has an affine image $\tilde{K}$ for which

$$\text{vol}(\partial \tilde{K}) \leq 2n \, \text{vol}(\tilde{K})^{(n-1)/n}.$$

Choose $\tilde{K}$ so that its maximal volume ellipsoid is $B_2^n$, the Euclidean ball of radius 1. The volume of $\tilde{K}$ is then at most $2^n$, since this is the volume of the cube whose maximal ellipsoid is $B_2^n$. As in the previous lecture,

$$\text{vol}(\partial \tilde{K}) = \lim_{\varepsilon \to 0} \frac{\text{vol}(\tilde{K} + \varepsilon B_2^n) - \text{vol}(\tilde{K})}{\varepsilon}.$$

Since $\tilde{K} \supset B_2^n$, the second expression is at most

$$\lim_{\varepsilon \to 0} \frac{\text{vol}(\tilde{K} + \varepsilon \tilde{K}) - \text{vol}(\tilde{K})}{\varepsilon} = \text{vol}(\tilde{K}) \lim_{\varepsilon \to 0} \frac{(1+\varepsilon)^n - 1}{\varepsilon}$$

$$= n \, \text{vol}(\tilde{K}) = n \, \text{vol}(\tilde{K})^{1/n} \, \text{vol}(\tilde{K})^{(n-1)/n}$$

$$\leq 2n \, \text{vol}(\tilde{K})^{(n-1)/n},$$

which is exactly what we wanted.

The rest of this lecture will thus be devoted to explaining the proof of the volume-ratio estimate:

THEOREM 6.2. *Among symmetric convex bodies the cube has largest volume ratio.*

As one might expect, the proof of Theorem 6.2 makes use of John's Theorem from Lecture 3. The problem is to show that, if $K$ is a convex body whose maximal ellipsoid is $B_2^n$, then $\text{vol}(K) \leq 2^n$. As we saw, it is a consequence of

John's theorem that if $B_2^n$ is the maximal ellipsoid in $K$, there is a sequence $(u_i)$ of unit vectors and a sequence $(c_i)$ of positive numbers for which

$$\sum c_i\, u_i \otimes u_i = I_n$$

and for which

$$K \subset C := \{x : |\langle x, u_i\rangle| \leq 1 \text{ for } 1 \leq i \leq m\}.$$

We shall show that this $C$ has volume at most $2^n$. The principal tool will be a sharp inequality for norms of generalised convolutions. Before stating this let's explain some standard terms from harmonic analysis.

If $f$ and $g : \mathbb{R} \to \mathbb{R}$ are bounded, integrable functions, we define the *convolution* $f * g$ of $f$ and $g$ by

$$f * g(x) = \int_{\mathbb{R}} f(y)g(x - y)\, dy.$$

Convolutions crop up in many areas of mathematics and physics, and a good deal is known about how they behave. One of the most fundamental inequalities for convolutions is Young's inequality: If $f \in L_p$, $g \in L_q$, and

$$\frac{1}{p} + \frac{1}{q} = 1 + \frac{1}{s},$$

then

$$\|f * g\|_s \leq \|f\|_p \|g\|_q.$$

(Here $\|\cdot\|_p$ means the $L_p$ norm on $\mathbb{R}$, and so on.) Once we have Young's inequality, we can give a meaning to convolutions of functions that are not both integrable and bounded, provided that they lie in the correct $L_p$ spaces. Young's inequality holds for convolution on any locally compact group, for example the circle. On compact groups it is sharp: there is equality for constant functions. But on $\mathbb{R}$, where constant functions are not integrable, the inequality can be improved (for most values of $p$ and $q$). It was shown by Beckner [1975] and Brascamp and Lieb [1976a] that the correct constant in Young's inequality is attained if $f$ and $g$ are appropriate Gaussian densities: that is, for some positive $a$ and $b$, $f(t) = e^{-at^2}$ and $g(t) = e^{-bt^2}$. (The appropriate choices of $a$ and $b$ and the value of the best constant for each $p$ and $q$ will not be stated here. Later we shall see that they can be avoided.)

How are convolutions related to convex bodies? To answer this question we need to rewrite Young's inequality slightly. If $1/r + 1/s = 1$, the $L_s$ norm $\|f * g\|_s$ can be realised as

$$\int_{\mathbb{R}} (f * g)(x)h(x)$$

for some function $h$ with $\|h\|_r = 1$. So the inequality says that, if $1/p+1/q+1/r = 2$, then

$$\iint f(y)g(x - y)h(x)\, dy\, dx \leq \|f\|_p \|g\|_q \|h\|_r.$$

We may rewrite the inequality again with $h(-x)$ in place of $h(x)$, since this doesn't affect $\|h\|_r$:

$$\iint f(y)g(x-y)h(-x)\,dy\,dx \leq \|f\|_p\,\|g\|_q\,\|h\|_r\,. \tag{6.1}$$

This can be written in a more symmetric form via the map from $\mathbb{R}^2$ into $\mathbb{R}^3$ that takes $(x,y)$ to $(y, x-y, -x) =: (u, v, w)$. The range of this map is the subspace

$$H = \{(u,v,w) : u+v+w = 0\}\,.$$

Apart from a factor coming from the Jacobian of this map, the integral can be written

$$\int_H f(u)g(v)h(w),$$

where the integral is with respect to two-dimensional measure on the subspace $H$. So Young's inequality and its sharp forms estimate the integral of a product function on $\mathbb{R}^3$ over a subspace. What is the simplest product function? If $f$, $g$, and $h$ are each the characteristic function of the interval $[-1,1]$, the function $F$ given by

$$F(u,v,w) = f(u)g(v)h(w)$$

is the characteristic function of the cube $[-1,1]^3 \subset \mathbb{R}^3$. The integral of $F$ over a subspace of $\mathbb{R}^3$ is thus the area of a slice of the cube: the area of a certain convex body. So there is some hope that we might use a convolution inequality to estimate volumes.

Brascamp and Lieb proved rather more than the sharp form of Young's inequality stated earlier. They considered not just two-dimensional subspaces of $\mathbb{R}^3$ but $n$-dimensional subspaces of $\mathbb{R}^m$. It will be more convenient to state their result using expressions analogous to those in (6.1) rather than using integrals over subspaces. Notice that the integral

$$\iint f(y)g(x-y)h(-x)\,dy\,dx$$

can be written

$$\int_{\mathbb{R}^2} f(\langle x, v_1\rangle)g(\langle x, v_2\rangle)h(\langle x, v_3\rangle)\,dx,$$

where $v_1 = (0,1)$, $v_2 = (1,-1)$ and $v_3 = (-1,0)$ are vectors in $\mathbb{R}^2$. The theorem of Brascamp and Lieb is the following.

THEOREM 6.3. *If $(v_i)_1^m$ are vectors in $\mathbb{R}^n$ and $(p_i)_1^m$ are positive numbers satisfying*

$$\sum_1^m \frac{1}{p_i} = n,$$

*and if $(f_i)_1^m$ are nonnegative measurable functions on the line, then*

$$\frac{\int_{\mathbb{R}^n} \prod_1^m f_i\left(\langle x, v_i\rangle\right)}{\prod_1^m \|f_i\|_{p_i}}$$

*is "maximised" when the $(f_i)$ are appropriate Gaussian densities: $f_i(t) = e^{-a_i t^2}$, where the $a_i$ depend upon $m$, $n$, the $p_i$, and the $v_i$.*

The word maximised is in quotation marks since there are degenerate cases for which the maximum is not attained. The value of the maximum is not easily computed since the $a_i$ are the solutions of nonlinear equations in the $p_i$ and $v_i$. This apparently unpleasant problem evaporates in the context of convex geometry: the inequality has a normalised form, introduced in [Ball 1990], which fits perfectly with John's Theorem.

THEOREM 6.4. *If $(u_i)_1^m$ are unit vectors in $\mathbb{R}^n$ and $(c_i)_1^m$ are positive numbers for which*

$$\sum_1^m c_i\, u_i \otimes u_i = I_n,$$

*and if $(f_i)_1^m$ are nonnegative measurable functions, then*

$$\int_{\mathbb{R}^n} \prod f_i\left(\langle x, u_i\rangle\right)^{c_i} \leq \prod \left(\int f_i\right)^{c_i}.$$

In this reformulation of the theorem, the $c_i$ play the role of $1/p_i$: the Fritz John condition ensures that $\sum c_i = n$ as required, and miraculously guarantees that the correct constant in the inequality is 1 (as written). The functions $f_i$ have been replaced by $f_i^{c_i}$, since this ensures that equality occurs if the $f_i$ are identical Gaussian densities. It may be helpful to see why this is so. If $f_i(t) = e^{-t^2}$ for all $i$, then

$$\prod f_i\left(\langle x, u_i\rangle\right)^{c_i} = \exp\left(-\sum c_i \langle x, u_i\rangle^2\right) = e^{-|x|^2} = \prod_1^n e^{-x_i^2},$$

so the integral is

$$\left(\int e^{-t^2}\right)^n = \prod\left(\int e^{-t^2}\right)^{c_i} = \prod\left(\int f_i\right)^{c_i}.$$

Armed with Theorem 6.4, let's now prove Theorem 6.2.

PROOF OF THE VOLUME-RATIO ESTIMATE. Recall that our aim is to show that, for $u_i$ and $c_i$ as usual, the body

$$C = \{x : |\langle x, u_i\rangle| \leq 1 \text{ for } 1 \leq i \leq m\}$$

has volume at most $2^n$. For each $i$ let $f_i$ be the characteristic function of the interval $[-1, 1]$ in $\mathbb{R}$. Then the function

$$x \mapsto \prod f_i(\langle x, u_i\rangle)^{c_i}$$

is exactly the characteristic function of $C$. Integrating and applying Theorem 6.4 we have

$$\operatorname{vol}(C) \leq \prod \left( \int f_i \right)^{c_i} = \prod 2^{c_i} = 2^n. \qquad \square$$

The theorems of Brascamp and Lieb and Beckner have been greatly extended over the last twenty years. The main result in Beckner's paper solved the old problem of determining the norm of the Fourier transform between $L_p$ spaces. There are many classical inequalities in harmonic analysis for which the best constants are now known. The paper [Lieb 1990] contains some of the most up-to-date discoveries and gives a survey of the history of these developments.

The methods described here have many other applications to convex geometry. There is also a reverse form of the Brascamp–Lieb inequality appropriate for analysing, for example, the ratio of the volume of a body to that of the minimal ellipsoid containing it.

## Lecture 7. The Central Limit Theorem and Large Deviation Inequalities

The material in this short lecture is not really convex geometry, but is intended to provide a context for what follows. For the sake of readers who may not be familiar with probability theory, we also include a few words about independent random variables.

To begin with, a *probability measure* $\mu$ on a set $\Omega$ is just a measure of total mass $\mu(\Omega) = 1$. Real-valued functions on $\Omega$ are called random variables and the integral of such a function $X : \Omega \to \mathbb{R}$, its mean, is written $EX$ and called the expectation of $X$. The variance of $X$ is $E(X - EX)^2$. It is customary to suppress the reference to $\Omega$ when writing the measures of sets defined by random variables. Thus

$$\mu(\{\omega \in \Omega : X(\omega) < 1\})$$

is written $\mu(X < 1)$: the probability that $X$ is less than 1.

Two crucial, and closely related, ideas distinguish probability theory from general measure theory. The first is independence. Two random variables $X$ and $Y$ are said to be independent if, for any functions $f$ and $g$,

$$Ef(X)g(Y) = Ef(X)\,Eg(Y).$$

Independence can always be viewed in a canonical way. Let $(\Omega, \mu)$ be a product space $(\Omega_1 \times \Omega_2, \mu_1 \otimes \mu_2)$, where $\mu_1$ and $\mu_2$ are probabilities. Suppose $X$ and $Y$ are random variables on $\Omega$ for which the value $X(\omega_1, \omega_2)$ depends only upon $\omega_1$ while $Y(\omega_1, \omega_2)$ depends only upon $\omega_2$. Then any integral (that converges appropriately)

$$Ef(X)g(Y) = \int f(X(s))\,g(Y(t))\,d\mu_1 \otimes \mu_2(s, t)$$

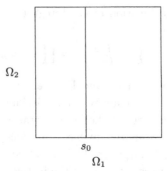

$\Omega_2$

$s_0$

$\Omega_1$

**Figure 23.** Independence and product spaces

can be written as the product of integrals

$$\int f(X(s)) \, d\mu_1(s) \int g(Y(t)) \, d\mu_2(t) = Ef(X)Eg(Y)$$

by Fubini's Theorem. Putting it another way, on each line $\{(s_0, t) : t \in \Omega_2\}$, $X$ is fixed, while $Y$ exhibits its full range of behaviour in the correct proportions. This is illustrated in Figure 23.

In a similar way, a sequence $X_1, X_2, \ldots, X_n$ of independent random variables arises if each variable is defined on the product space $\Omega_1 \times \Omega_2 \times \ldots \times \Omega_n$ and $X_i$ depends only upon the $i$-th coordinate.

The second crucial idea, which we will not discuss in any depth, is the use of many different $\sigma$-fields on the same space. The simplest example has already been touched upon. The product space $\Omega_1 \times \Omega_2$ carries two $\sigma$-fields, much smaller than the product field, which it inherits from $\Omega_1$ and $\Omega_2$ respectively. If $\mathcal{F}_1$ and $\mathcal{F}_2$ are the $\sigma$-fields on $\Omega_1$ and $\Omega_2$, the sets of the form $A \times \Omega_2 \subset \Omega_1 \times \Omega_2$ for $A \in \mathcal{F}_1$ form a $\sigma$-field on $\Omega_1 \times \Omega_2$; let's call it $\tilde{\mathcal{F}}_1$. Similarly,

$$\tilde{\mathcal{F}}_2 = \{\Omega_1 \times B : B \in \mathcal{F}_2\}.$$

"Typical" members of these $\sigma$-fields are shown in Figure 24.

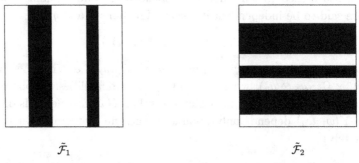

$\tilde{\mathcal{F}}_1$                                     $\tilde{\mathcal{F}}_2$

**Figure 24.** Members of the "small" $\sigma$-fields $\tilde{\mathcal{F}}_1$ and $\tilde{\mathcal{F}}_2$ on $\Omega_1 \times \Omega_2$.

One of the most beautiful and significant principles in mathematics is the central limit theorem: any random quantity that arises as the sum of many small independent contributions is distributed very much like a Gaussian random variable. The most familiar example is coin tossing. We use a coin whose decoration is a bit austere: it has $+1$ on one side and $-1$ on the other. Let $\varepsilon_1, \varepsilon_2, \ldots, \varepsilon_n$ be the outcomes of $n$ independent tosses. Thus the $\varepsilon_i$ are independent random variables, each of which takes the values $+1$ and $-1$ each with probability $\frac{1}{2}$. (Such random variables are said to have a *Bernoulli distribution*.) Then the normalised sum

$$S_n = \frac{1}{\sqrt{n}} \sum_1^n \varepsilon_i$$

belongs to an interval $I$ of the real line with probability very close to

$$\frac{1}{\sqrt{2\pi}} \int_I e^{-t^2/2} \, dt.$$

The normalisation $1/\sqrt{n}$, ensures that the variance of $S_n$ is 1: so there is some hope that the $S_n$ will all be similarly distributed.

The standard proof of the central limit theorem shows that much more is true. Any sum of the form

$$\sum_1^n a_i \varepsilon_i$$

with real coefficients $a_i$ will have a roughly Gaussian distribution as long as each $a_i$ is fairly small compared with $\sum a_i^2$. Some such smallness condition is clearly needed since if

$$a_1 = 1 \quad \text{and} \quad a_2 = a_3 = \cdots = a_n = 0,$$

the sum is just $\varepsilon_1$, which is not much like a Gaussian. However, in many instances, what one really wants is not that the sum is distributed like a Gaussian, but merely that the sum cannot be large (or far from average) much more often than an appropriate Gaussian variable. The example above clearly satisfies such a condition: $\varepsilon_1$ never deviates from its mean, 0, by more than 1.

The following inequality provides a deviation estimate for any sequence of coefficients. In keeping with the custom among functional analysts, I shall refer to the inequality as Bernstein's inequality. (It is not related to the Bernstein inequality for polynomials on the circle.) However, probabilists know the result as Hoeffding's inequality, and the earliest reference known to me is [Hoeffding 1963]. A stronger and more general result goes by the name of the Azuma–Hoeffding inequality; see [Williams 1991], for example.

THEOREM 7.1 (BERNSTEIN'S INEQUALITY). *If $\varepsilon_1, \varepsilon_2, \ldots, \varepsilon_n$ are independent Bernoulli random variables and if $a_1, a_2, \ldots, a_n$ satisfy $\sum a_i^2 = 1$, then for each positive $t$ we have*

$$\text{Prob}\left( \left| \sum_{i=1}^n a_i \varepsilon_i \right| > t \right) \le 2e^{-t^2/2}.$$

This estimate compares well with the probability of finding a standard Gaussian outside the interval $[-t, t]$,

$$\frac{2}{\sqrt{2\pi}} \int_t^\infty e^{-s^2/2} \, ds.$$

The method by which Bernstein's inequality is proved has become an industry standard.

PROOF. We start by showing that, for each real $\lambda$,

$$Ee^{\lambda \sum a_i \varepsilon_i} \leq e^{\lambda^2/2}. \tag{7.1}$$

The idea will then be that $\sum a_i \varepsilon_i$ cannot be large too often, since, whenever it is large, its exponential is enormous.

To prove (7.1), we write

$$Ee^{\lambda \sum a_i \varepsilon_i} = E \prod_1^n e^{\lambda a_i \varepsilon_i}$$

and use independence to deduce that this equals

$$\prod_1^n Ee^{\lambda a_i \varepsilon_i}.$$

For each $i$ the expectation is

$$Ee^{\lambda a_i \varepsilon_i} = \frac{e^{\lambda a_i} + e^{-\lambda a_i}}{2} = \cosh \lambda a_i.$$

Now, $\cosh x \leq e^{x^2/2}$ for any real $x$, so, for each $i$,

$$Ee^{\lambda a_i \varepsilon_i} \leq e^{\lambda^2 a_i^2/2}.$$

Hence

$$Ee^{\lambda \sum a_i \varepsilon_i} \leq \prod_1^n e^{\lambda^2 a_i^2/2} = e^{\lambda^2/2},$$

since $\sum a_i^2 = 1$.

To pass from (7.1) to a probability estimate, we use the inequality variously known as Markov's or Chebyshev's inequality: if $X$ is a nonnegative random variable and $R$ is positive, then

$$R \operatorname{Prob}(X \geq R) \leq EX$$

(because the integral includes a bit where a function whose value is at least $R$ is integrated over a set of measure $\operatorname{Prob}(X \geq R)$).

Suppose $t \geq 0$. Whenever $\sum a_i \varepsilon_i \geq t$, we will have $e^{t \sum a_i \varepsilon_i} \geq e^{t^2}$. Hence

$$e^{t^2} \operatorname{Prob}\left(\sum a_i \varepsilon_i \geq t\right) \leq Ee^{t \sum a_i \varepsilon_i} \leq e^{t^2/2}$$

by (7.1). So

$$\text{Prob}\left(\sum a_i \varepsilon_i \geq t\right) \leq e^{-t^2/2},$$

and in a similar way we get

$$\text{Prob}\left(\sum a_i \varepsilon_i \leq -t\right) \leq e^{-t^2/2}.$$

Putting these inequalities together we get

$$\text{Prob}\left(\left|\sum a_i \varepsilon_i\right| \geq t\right) \leq 2e^{-t^2/2}. \qquad \square$$

In the next lecture we shall see that deviation estimates that look like Bernstein's inequality hold for a wide class of functions in several geometric settings. For the moment let's just remark that an estimate similar to Bernstein's inequality,

$$\text{Prob}\left(\left|\sum a_i X_i\right| \geq t\right) \leq 2e^{-6t^2}, \qquad (7.2)$$

holds for $\sum a_i^2 = 1$, if the $\pm 1$ valued random variables $\varepsilon_i$ are replaced by independent random variables $X_i$ each of which is uniformly distributed on the interval $\left[-\frac{1}{2}, \frac{1}{2}\right]$. This already has a more geometric flavour, since for these $(X_i)$ the vector $(X_1, X_2, \ldots, X_n)$ is distributed according to Lebesgue measure on the cube $\left[-\frac{1}{2}, \frac{1}{2}\right]^n \subset \mathbb{R}^n$. If $\sum a_i^2 = 1$, then $\sum a_i X_i$ is the distance of the point $(X_1, X_2, \ldots, X_n)$ from the subspace of $\mathbb{R}^n$ orthogonal to $(a_1, a_2, \ldots, a_n)$. So (7.2) says that most of the mass of the cube lies close to any subspace of $\mathbb{R}^n$, which is reminiscent of the situation for the Euclidean ball described in Lecture 1.

## Lecture 8. Concentration of Measure in Geometry

The aim of this lecture is to describe geometric analogues of Bernstein's deviation inequality. These geometric deviation estimates are closely related to isoperimetric inequalities. The phenomenon of which they form a part was introduced into the field by V. Milman: its development, especially by Milman himself, led to a new, probabilistic, understanding of the structure of convex bodies in high dimensions. The phenomenon was aptly named the concentration of measure.

We explained in Lecture 5 how the Brunn–Minkowski inequality implies the classical isoperimetric inequality in $\mathbb{R}^n$: among bodies of a given volume, the Euclidean balls have least surface area. There are many other situations where isoperimetric inequalities are known; two of them will be described below. First let's recall that the argument from the Brunn–Minkowski inequality shows more than the isoperimetric inequality.

Let $A$ be a compact subset of $\mathbb{R}^n$. For each point $x$ of $\mathbb{R}^n$, let $d(x, A)$ be the distance from $x$ to $A$:

$$d(x, A) = \min\left\{|x - y| : y \in A\right\}.$$

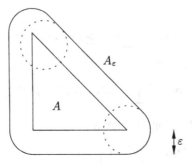

**Figure 25.** An $\varepsilon$-neighbourhood.

For each positive $\varepsilon$, the Minkowski sum $A + \varepsilon B_2^n$ is exactly the set of points whose distance from $A$ is at most $\varepsilon$. Let's denote such an $\varepsilon$-neighbourhood $A_\varepsilon$; see Figure 25.

The Brunn–Minkowski inequality shows that, if $B$ is an Euclidean ball of the same volume as $A$, we have

$$\operatorname{vol}(A_\varepsilon) \geq \operatorname{vol}(B_\varepsilon) \quad \text{for any } \varepsilon > 0.$$

This formulation of the isoperimetric inequality makes much clearer the fact that it relates the measure and the metric on $\mathbb{R}^n$. If we blow up a set in $\mathbb{R}^n$ using the metric, we increase the measure by at least as much as we would for a ball.

This idea of comparing the volumes of a set and its neighbourhoods makes sense in any space that has both a measure and a metric, regardless of whether there is an analogue of Minkowski addition. For any metric space $(\Omega, d)$ equipped with a Borel measure $\mu$, and any positive $\alpha$ and $\varepsilon$, it makes sense to ask: For which sets $A$ of measure $\alpha$ do the blow-ups $A_\varepsilon$ have smallest measure? This general isoperimetric problem has been solved in a variety of different situations. We shall consider two closely related geometric examples. In each case the measure $\mu$ will be a probability measure: as we shall see, in this case, isoperimetric inequalities may have totally unexpected consequences.

In the first example, $\Omega$ will be the sphere $S^{n-1}$ in $\mathbb{R}^n$, equipped with either the geodesic distance or, more simply, the Euclidean distance inherited from $\mathbb{R}^n$ as shown in Figure 26. (This is also called the *chordal metric*; it was used in Lecture 2 when we discussed spherical caps of given radii.) The measure will be $\sigma = \sigma_{n-1}$, the rotation-invariant probability on $S^{n-1}$. The solutions of the isoperimetric problem on the sphere are known exactly: they are spherical caps (Figure 26, right) or, equivalently, they are balls in the metric on $S^{n-1}$. Thus, if a subset $A$ of the sphere has the same measure as a cap of radius $r$, its neighbourhood $A_\varepsilon$ has measure at least that of a cap of radius $r + \varepsilon$.

This statement is a good deal more difficult to prove than the classical isoperimetric inequality on $\mathbb{R}^n$: it was discovered by P. Lévy, quite some time after the isoperimetric inequality in $\mathbb{R}^n$. At first sight, the statement looks innocuous

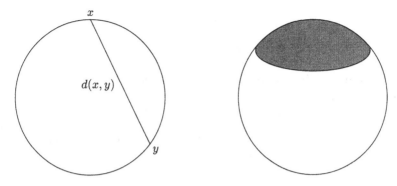

**Figure 26.** The Euclidean metric on the sphere. A spherical cap (right) is a ball for this metric.

enough (despite its difficulty): but it has a startling consequence. Suppose $\alpha = \frac{1}{2}$, so that $A$ has the measure of a hemisphere $H$. Then, for each positive $\varepsilon$, the set $A_\varepsilon$ has measure at least that of the set $H_\varepsilon$, illustrated in Figure 27. The complement of $H_\varepsilon$ is a spherical cap that, as we saw in Lecture 2, has measure about $e^{-n\varepsilon^2/2}$. Hence $\sigma(A_\varepsilon) \geq 1 - e^{-n\varepsilon^2/2}$, so almost the entire sphere lies within distance $\varepsilon$ of $A$, even though there may be points rather far from $A$. The measure and the metric on the sphere "don't match": the mass of $\sigma$ concentrates very close to any set of measure $\frac{1}{2}$. This is clearly related to the situation described in Lecture 1, in which we found most of the mass of the ball concentrated near each hyperplane: but now the phenomenon occurs for any set of measure $\frac{1}{2}$.

The phenomenon just described becomes even more striking when reinterpreted in terms of Lipschitz functions. Suppose $f : S^{n-1} \to \mathbb{R}$ is a function on the sphere that is 1-Lipschitz: that is, for any pair of points $\theta$ and $\phi$ on the sphere,

$$|f(\theta) - f(\phi)| \leq |\theta - \phi|.$$

There is at least one number $M$, the median of $f$, for which both the sets $(f \leq M)$ and $(f \geq M)$ have measure at least $\frac{1}{2}$. If a point $x$ has distance at most $\varepsilon$ from

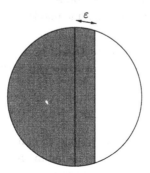

**Figure 27.** An $\varepsilon$-neighbourhood of a hemisphere.

$(f \leq M)$, then (since $f$ is 1-Lipschitz)

$$f(x) \leq M + \varepsilon.$$

By the isoperimetric inequality all but a tiny fraction of the points on the sphere have this property:

$$\sigma(f > M + \varepsilon) \leq e^{-n\varepsilon^2/2}.$$

Similarly, $f$ is larger than $M - \varepsilon$ on all but a fraction of the sphere. Putting these statements together we get

$$\sigma(|f - M| > \varepsilon) \leq 2e^{-n\varepsilon^2/2}.$$

So, although $f$ may vary by as much as 2 between a point of the sphere and its opposite, the function is nearly equal to $M$ on almost the entire sphere: $f$ is practically constant.

In the case of the sphere we thus have the following pair of properties.

(i) If $A \subset \Omega$ with $\mu(A) = \frac{1}{2}$ then $\mu(A_\varepsilon) \geq 1 - e^{-n\varepsilon^2/2}$.
(ii) If $f : \Omega \to \mathbb{R}$ is 1-Lipschitz there is a number $M$ for which

$$\mu(|f - M| > \varepsilon) \leq 2e^{-n\varepsilon^2/2}.$$

Each of these statements may be called an approximate isoperimetric inequality. We have seen how the second can be deduced from the first. The reverse implication also holds (apart from the precise constants involved). (To see why, apply the second property to the function given by $f(x) = d(x, A)$.)

In many applications, exact solutions of the isoperimetric problem are not as important as deviation estimates of the kind we are discussing. In some cases where the exact solutions are known, the two properties above are a good deal easier to prove than the solutions: and in a great many situations, an exact isoperimetric inequality is not known, but the two properties are. The formal similarity between property 2 and Bernstein's inequality of the last lecture is readily apparent. There are ways to make this similarity much more than merely formal: there are deviation inequalities that have implications for Lipschitz functions and imply Bernstein's inequality, but we shall not go into them here.

In our second example, the space $\Omega$ will be $\mathbb{R}^n$ equipped with the ordinary Euclidean distance. The measure will be the standard Gaussian probability measure on $\mathbb{R}^n$ with density

$$\gamma(x) = (2\pi)^{-n/2} e^{-|x|^2/2}.$$

The solutions of the isoperimetric problem in Gauss space were found by Borell [1975]. They are half-spaces. So, in particular, if $A \subset \mathbb{R}^n$ and $\mu(A) = \frac{1}{2}$, then $\mu(A_\varepsilon)$ is at least as large as $\mu(H_\varepsilon)$, where $H$ is the half-space $\{x \in \mathbb{R}^n : x_1 \leq 0\}$ and so $H_\varepsilon = \{x : x_1 \leq \varepsilon\}$: see Figure 28.

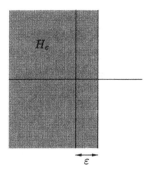

**Figure 28.** An $\varepsilon$-neighbourhood of a half-space.

The complement of $H_\varepsilon$ has measure

$$\frac{1}{\sqrt{2\pi}} \int_\varepsilon^\infty e^{-t^2/2}\, dt \le e^{-\varepsilon^2/2}.$$

Hence,

$$\mu(A_\varepsilon) \ge 1 - e^{-\varepsilon^2/2}.$$

Since $n$ does not appear in the exponent, this looks much weaker than the statement for the sphere, but we shall see that the two are more or less equivalent.

Borell proved his inequality by using the isoperimetric inequality on the sphere. A more direct proof of a deviation estimate like the one just derived was found by Maurey and Pisier, and their argument gives a slightly stronger, Sobolev-type inequality [Pisier 1989, Chapter 4]. We too shall aim directly for a deviation estimate, but a little background to the proof may be useful.

There was an enormous growth in understanding of approximate isoperimetric inequalities during the late 1980s, associated most especially with the name of Talagrand. The reader whose interest has been piqued should certainly look at Talagrand's gorgeous articles [1988; 1991a], in which he describes an approach to deviation inequalities in product spaces that involves astonishingly few structural hypotheses. In a somewhat different vein (but prompted by his earlier work), Talagrand [1991b] also found a general principle, strengthening the approximate isoperimetric inequality in Gauss space. A simplification of this argument was found by Maurey [1991]. The upshot is that a deviation inequality for Gauss space can be proved with an extremely short argument that fits naturally into these notes.

THEOREM 8.1 (APPROXIMATE ISOPERIMETRIC INEQUALITY FOR GAUSS SPACE).
*Let $A \subset \mathbb{R}^n$ be measurable and let $\mu$ be the standard Gaussian measure on $\mathbb{R}^n$. Then*

$$\int e^{d(x,A)^2/4}\, d\mu \le \frac{1}{\mu(A)}.$$

*Consequently, if $\mu(A) = \frac{1}{2}$,*

$$\mu(A_\varepsilon) \ge 1 - 2e^{-\varepsilon^2/4}.$$

PROOF. We shall deduce the first assertion directly from the Prékopa–Leindler inequality (with $\lambda = \frac{1}{2}$) of Lecture 5. To this end, define functions $f$, $g$, and $m$ on $\mathbb{R}^n$, as follows:

$$f(x) = e^{d(x,A)^2/4}\,\gamma(x),$$
$$g(x) = \chi_A(x)\,\gamma(x),$$
$$m(x) = \gamma(x),$$

where $\gamma$ is the Gaussian density. The assertion to be proved is that

$$\left(\int e^{d(x,A)^2/4}\,d\mu\right)\mu(A) \leq 1,$$

which translates directly into the inequality

$$\left(\int_{\mathbb{R}^n} f\right)\left(\int_{\mathbb{R}^n} g\right) \leq \left(\int_{\mathbb{R}^n} m\right)^2.$$

By the Prékopa–Leindler inequality it is enough to check that, for any $x$ and $y$ in $\mathbb{R}^n$,

$$f(x)g(y) \leq m\left(\frac{x+y}{2}\right)^2.$$

It suffices to check this for $y \in A$, since otherwise $g(y) = 0$. But, in this case, $d(x, A) \leq |x - y|$. Hence

$$(2\pi)^n f(x)g(y) = e^{d(x,A)^2/4}e^{-x^2/2}e^{-y^2/2}$$
$$\leq \exp\left(\frac{|x-y|^2}{4} - \frac{|x|^2}{2} - \frac{|y|^2}{2}\right) = \exp\left(-\frac{|x+y|^2}{4}\right)$$
$$= \left(\exp\left(-\frac{1}{2}\left|\frac{x+y}{2}\right|^2\right)\right)^2 = (2\pi)^n\, m\left(\frac{x+y}{2}\right)^2,$$

which is what we need.

To deduce the second assertion from the first, we use Markov's inequality, very much as in the proof of Bernstein's inequality of the last lecture. If $\mu(A) = \frac{1}{2}$, then

$$\int e^{d(x,A)^2/4}\,d\mu \leq 2.$$

The integral is at least

$$e^{\varepsilon^2/4}\mu\big(d(x, A) \geq \varepsilon\big).$$

So

$$\mu\big(d(x, A) \geq \varepsilon\big) \leq 2e^{-\varepsilon^2/4},$$

and the assertion follows.                                                        □

It was mentioned earlier that the Gaussian deviation estimate above is essentially equivalent to the concentration of measure on $S^{n-1}$. This equivalence depends upon the fact that the Gaussian measure in $\mathbb{R}^n$ is concentrated in a spherical shell of thickness approximately 1, and radius approximately $\sqrt{n}$. (Recall that the Euclidean ball of volume 1 has radius approximately $\sqrt{n}$.) This concentration is easily checked by direct computation using integration in spherical polars: but the inequality we just proved will do the job instead. There is an Euclidean ball of some radius $R$ whose Gaussian measure is $\frac{1}{2}$. According to the theorem above, Gaussian measure concentrates near the boundary of this ball. It is not hard to check that $R$ is about $\sqrt{n}$. This makes it quite easy to show that the deviation estimate for Gaussian measure guarantees a deviation estimate on the sphere of radius $\sqrt{n}$ with a decay rate of about $e^{-\varepsilon^2/4}$. If everything is scaled down by a factor of $\sqrt{n}$, onto the sphere of radius 1, we get a deviation estimate that decays like $e^{-n\varepsilon^2/4}$ and $n$ now appears in the exponent. The details are left to the reader.

The reader will probably have noticed that these estimates for Gauss space and the sphere are not quite as strong as those advertised earlier, because in each case the exponent is $\ldots \varepsilon^2/4 \ldots$ instead of $\ldots \varepsilon^2/2 \ldots$. In some applications, the sharper results are important, but for our purposes the difference will be irrelevant. It was pointed out to me by Talagrand that one can get as close as one wishes to the correct exponent $\ldots \varepsilon^2/2 \ldots$ by using the Prékopa–Leindler inequality with $\lambda$ close to 1 instead of $\frac{1}{2}$ and applying it to slightly different $f$ and $g$.

For the purposes of the next lecture we shall assume an estimate of $e^{-\varepsilon^2/2}$, even though we proved a weaker estimate.

## Lecture 9. Dvoretzky's Theorem

Although this is the ninth lecture, its subject, Dvoretzky's Theorem, was really the start of the modern theory of convex geometry in high dimensions. The phrase "Dvoretzky's Theorem" has become a generic term for statements to the effect that high-dimensional bodies have almost ellipsoidal slices. Dvoretzky's original proof shows that any symmetric convex body in $\mathbb{R}^n$ has almost ellipsoidal sections of dimension about $\sqrt{\log n}$. A few years after the publication of Dvoretzky's work, Milman [Milman 1971] found a very different proof, based upon the concentration of measure, which gave slices of dimension $\log n$. As we saw in Lecture 2 this is the best one can get in general. Milman's argument gives the following.

THEOREM 9.1. *There is a positive number $c$ such that, for every $\varepsilon > 0$ and every natural number $n$, every symmetric convex body of dimension $n$ has a slice of dimension*

$$k \geq \frac{c\varepsilon^2}{\log(1 + \varepsilon^{-1})} \log n$$

*that is within distance $1 + \varepsilon$ of the k-dimensional Euclidean ball.*

There have been many other proofs of similar results over the years. A particularly elegant one [Gordon 1985] gives the estimate $k \geq c\varepsilon^2 \log n$ (removing the logarithmic factor in $\varepsilon^{-1}$), and this estimate is essentially best possible. We chose to describe Milman's proof because it is conceptually easier to motivate and because the concentration of measure has many other uses. A few years ago, Schechtman found a way to eliminate the log factor within this approach, but we shall not introduce this subtlety here. We shall also not make any effort to be precise about the dependence upon $\varepsilon$.

With the material of Lecture 8 at our disposal, the plan of proof of Theorem 9.1 is easy to describe. We start with a symmetric convex body and we consider a linear image $K$ whose maximal volume ellipsoid is the Euclidean ball. For this $K$ we will try to find almost spherical sections, rather than merely ellipsoidal ones. Let $\|\cdot\|$ be the norm on $\mathbb{R}^n$ whose unit ball is $K$. We are looking for a $k$-dimensional space $H$ with the property that the function

$$\theta \mapsto \|\theta\|$$

is almost constant on the Euclidean sphere of $H$, $H \cap S^{n-1}$. Since $K$ contains $B_2^n$, we have $\|x\| \leq |x|$ for all $x \in \mathbb{R}^n$, so for any $\theta$ and $\phi$ in $S^{n-1}$,

$$|\|\theta\| - \|\phi\|| \leq \|\theta - \phi\| \leq |\theta - \phi|.$$

Thus $\|\cdot\|$ is a Lipschitz function on the sphere in $\mathbb{R}^n$, (indeed on all of $\mathbb{R}^n$). (We used the same idea in Lecture 4.) From Lecture 8 we conclude that the value of $\|\theta\|$ is almost constant on a very large proportion of $S^{n-1}$: it is almost equal to its average

$$M = \int_{S^{n-1}} \|\theta\| \, d\sigma,$$

on most of $S^{n-1}$.

We now choose our $k$-dimensional subspace at random. (The exact way to do this will be described below.) We can view this as a random embedding

$$T : \mathbb{R}^k \to \mathbb{R}^n.$$

For any particular unit vector $\psi \in \mathbb{R}^k$, there is a very high probability that its image $T\psi$ will have norm $\|T\psi\|$ close to $M$. This means that even if we select quite a number of vectors $\psi_1, \psi_2, \ldots, \psi_m$ in $S^{k-1}$ we can guarantee that there will be some choice of $T$ for which *all* the norms $\|T\psi_i\|$ will be close to $M$. We will thus have managed to pin down the radius of our slice in many different directions. If we are careful to distribute these directions well over the sphere in $\mathbb{R}^k$, we may hope that the radius will be almost constant on the entire sphere. For these purposes, "well distributed" will mean that all points of the sphere in $\mathbb{R}^k$ are close to one of our chosen directions. As in Lecture 2 we say that a set $\{\psi_1, \psi_2, \ldots, \psi_m\}$ in $S^{k-1}$ is a $\delta$-net for the sphere if every point of $S^{k-1}$ is within

(Euclidean) distance $\delta$ of at least one $\psi_i$. The arguments in Lecture 2 show that $S^{k-1}$ has a $\delta$-net with no more than

$$m = \left(\frac{4}{\delta}\right)^k$$

elements. The following lemma states that, indeed, pinning down the norm on a very fine net, pins it down everywhere.

LEMMA 9.2. *Let* $\|\cdot\|$ *be a norm on* $\mathbb{R}^k$ *and suppose that for each point* $\psi$ *of some* $\delta$-net on $S^{k-1}$, *we have*

$$M(1 - \gamma) \le \|\psi\| \le M(1 + \gamma)$$

*for some* $\gamma > 0$. *Then, for every* $\theta \in S^{k-1}$,

$$\frac{M(1 - \gamma - 2\delta)}{1 - \delta} \le \|\theta\| \le \frac{M(1 + \gamma)}{1 - \delta}.$$

PROOF. Clearly the value of $M$ plays no real role here so assume it is 1. We start with the upper bound. Let $C$ be the maximum possible ratio $\|x\|/|x|$ for nonzero $x$ and let $\theta$ be a point of $S^{k-1}$ with $\|\theta\| = C$. Choose $\psi$ in the $\delta$-net with $|\theta - \psi| \le \delta$. Then $\|\theta - \psi\| \le C|\theta - \psi| \le C\delta$, so

$$C = \|\theta\| \le \|\psi\| + \|\theta - \psi\| \le (1 + \gamma) + C\delta.$$

Hence

$$C \le \frac{(1 + \gamma)}{1 - \delta}.$$

To get the lower bound, pick some $\theta$ in the sphere and some $\psi$ in the $\delta$-net with $|\psi - \theta| \le \delta$. Then

$$(1 - \gamma) \le \|\psi\| \le \|\theta\| + \|\psi - \theta\| \le \|\theta\| + \frac{(1 + \gamma)}{1 - \delta}|\psi - \theta| \le \|\theta\| + \frac{(1 + \gamma)\delta}{1 - \delta}.$$

Hence

$$\|\theta\| \ge \left(1 - \gamma - \frac{\delta(1 + \gamma)}{1 - \delta}\right) = \frac{(1 - \gamma - 2\delta)}{1 - \delta}. \qquad \square$$

According to the lemma, our approach will give us a slice that is within distance

$$\frac{1 + \gamma}{1 - \gamma - 2\delta}$$

of the Euclidean ball (provided we satisfy the hypotheses), and this distance can be made as close as we wish to 1 if $\gamma$ and $\delta$ are small enough.

We are now in a position to prove the basic estimate.

THEOREM 9.3. *Let* $K$ *be a symmetric convex body in* $\mathbb{R}^n$ *whose ellipsoid of maximal volume is* $B_2^n$ *and put*

$$M = \int_{S^{n-1}} \|\theta\| \, d\sigma$$

*as above. Then $K$ has almost spherical slices whose dimension is of the order of $nM^2$.*

PROOF. Choose $\gamma$ and $\delta$ small enough to give the desired accuracy, in accordance with the lemma.

Since the function $\theta \mapsto \|\theta\|$ is Lipschitz (with constant 1) on $S^{n-1}$, we know from Lecture 8 that, for any $t \geq 0$,

$$\sigma\big(|\|\theta\| - M| > t\big) \leq 2e^{-nt^2/2}.$$

In particular,

$$\sigma\big(|\|\theta\| - M| > M\gamma\big) \leq 2e^{-nM^2\gamma^2/2}.$$

So

$$M(1 - \gamma) \leq \|\theta\| \leq M(1 + \gamma)$$

on all but a proportion $2e^{-nM^2\gamma^2/2}$ of the sphere.

Let $\mathcal{A}$ be a $\delta$-net on the sphere in $\mathbb{R}^k$ with at most $(4/\delta)^k$ elements. Choose a random embedding of $\mathbb{R}^k$ in $\mathbb{R}^n$: more precisely, fix a particular copy of $\mathbb{R}^k$ in $\mathbb{R}^n$ and consider its images under orthogonal transformations $U$ of $\mathbb{R}^n$ as a random subspace with respect to the invariant probability on the group of orthogonal transformations. For each fixed $\psi$ in the sphere of $\mathbb{R}^k$, its images $U\psi$, are uniformly distributed on the sphere in $\mathbb{R}^n$. So for each $\psi \in \mathcal{A}$, the inequality

$$M(1 - \gamma) \leq \|U\psi\| \leq M(1 + \gamma)$$

holds for $U$ outside a set of measure at most $2e^{-nM^2\gamma^2/2}$. So there will be at least one $U$ for which this inequality holds for *all* $\psi$ in $\mathcal{A}$, as long as the sum of the probabilities of the bad sets is at most 1. This is guaranteed if

$$\left(\frac{4}{\delta}\right)^k 2e^{-nM^2\gamma^2/2} < 1.$$

This inequality is satisfied by $k$ of the order of

$$nM^2 \frac{\gamma^2}{2\log(4/\delta)}. \qquad \qquad \square$$

Theorem 9.3 guarantees the existence of spherical slices of $K$ of large dimension, provided the average

$$M = \int_{S^{n-1}} \|\theta\| \, d\sigma$$

is not too small. Notice that we certainly have $M \leq 1$ since $\|x\| \leq |x|$ for all $x$. In order to get Theorem 9.1 from Theorem 9.3 we need to get a lower estimate for $M$ of the order of

$$\frac{\sqrt{\log n}}{\sqrt{n}}.$$

This is where we must use the fact that $B_2^n$ is the *maximal* volume ellipsoid in $K$. We saw in Lecture 3 that in this situation $K \subset \sqrt{n}B_2^n$, so $\|x\| \geq |x|/\sqrt{n}$ for all $x$, and hence

$$M \geq \frac{1}{\sqrt{n}}.$$

But this estimate is useless, since it would not give slices of dimension bigger than 1. It is vital that we use the more detailed information provided by John's Theorem.

Before we explain how this works, let's look at our favourite examples. For specific norms it is usually much easier to compute the mean $M$ by writing it as an integral with respect to Gaussian measure on $\mathbb{R}^n$. As in Lecture 8 let $\mu$ be the standard Gaussian measure on $\mathbb{R}^n$, with density

$$(2\pi)^{-n/2}e^{-|x|^2/2}.$$

By using polar coordinates we can write

$$\int_{S^{n-1}} \|\theta\|\, d\sigma = \frac{\Gamma(n/2)}{\sqrt{2}\Gamma((n+1)/2)} \int_{\mathbb{R}^n} \|x\|\, d\mu(x) > \frac{1}{\sqrt{n}} \int_{\mathbb{R}^n} \|x\|\, d\mu(x).$$

The simplest norm for which to calculate is the $\ell_1$ norm. Since the body we consider is supposed to have $B_2^n$ as its maximal ellipsoid we must use $\sqrt{n}B_1^n$, for which the corresponding norm is

$$\|x\| = \frac{1}{\sqrt{n}} \sum_1^n |x_i|.$$

Since the integral of this sum is just $n$ times the integral of any one coordinate it is easy to check that

$$\frac{1}{\sqrt{n}} \int_{\mathbb{R}^n} \|x\|\, d\mu(x) = \sqrt{\frac{2}{\pi}}.$$

So for the scaled copies of $B_1^n$, we have $M$ bounded below by a fixed number, and Theorem 9.3 guarantees almost spherical sections of dimension proportional to $n$. This was first proved, using exactly the method described here, in [Figiel et al. 1977], which had a tremendous influence on subsequent developments. Notice that this result and Kašin's Theorem from Lecture 4 are very much in the same spirit, but neither implies the other. The method used here does not achieve dimensions as high as $n/2$ even if we are prepared to allow quite a large distance from the Euclidean ball. On the other hand, the volume-ratio argument does not give sections that are very close to Euclidean: the volume ratio is the closest one gets this way. Some time after Kašin's article appeared, the gap between these results was completely filled by Garnaev and Gluskin [Garnaev and Gluskin 1984]. An analogous gap in the general setting of Theorem 9.1, namely that the existing proofs could not give a dimension larger than some fixed multiple of $\log n$, was recently filled by Milman and Schechtman.

What about the cube? This body *has* $B_2^n$ as its maximal ellipsoid, so our job is to estimate

$$\frac{1}{\sqrt{n}} \int_{\mathbb{R}^n} \max |x_i| \, d\mu(x).$$

At first sight this looks much more complicated than the calculation for $B_1^n$, since we cannot simplify the integral of a maximum. But, instead of estimating the mean of the function $\max |x_i|$, we can estimate its median (and from Lecture 8 we know that they are not far apart). So let $R$ be the number for which

$$\mu \left( \max |x_i| \leq R \right) = \mu \left( \max |x_i| \geq R \right) = \tfrac{1}{2}.$$

From the second identity we get

$$\frac{1}{\sqrt{n}} \int_{\mathbb{R}^n} \max |x_i| \, d\mu(x) \geq \frac{R}{2\sqrt{n}}.$$

We estimate $R$ from the first identity. It says that the cube $[-R, R]^n$ has Gaussian measure $\tfrac{1}{2}$. But the cube is a "product" so

$$\mu \left( [-R, R]^n \right) = \left( \frac{1}{\sqrt{2\pi}} \int_{-R}^{R} e^{-t^2/2} \, dt \right)^n.$$

In order for this to be equal to $\tfrac{1}{2}$ we need the expression

$$\frac{1}{\sqrt{2\pi}} \int_{-R}^{R} e^{-t^2/2} \, dt$$

to be about $1 - (\log 2)/n$. Since the expression approaches 1 roughly like

$$1 - e^{-R^2/2},$$

we get an estimate for $R$ of the order of $\sqrt{\log n}$. From Theorem 9.3 we then recover the simple result of Lecture 2 that the cube has almost spherical sections of dimension about $\log n$.

There are many other bodies and classes of bodies for which $M$ can be efficiently estimated. For example, the correct order of the largest dimension of Euclidean slice of the $\ell_p^n$ balls, was also computed in the paper [Figiel et al. 1977] mentioned earlier.

We would like to know that for a *general* body with maximal ellipsoid $B_2^n$ we have

$$\int_{\mathbb{R}^n} \|x\| \, d\mu(x) \geq (\text{constant}) \sqrt{\log n} \tag{9.1}$$

just as we do for the cube. The usual proof of this goes via the Dvoretzky–Rogers Lemma, which can be proved using John's Theorem. This is done for example in [Pisier 1989]. Roughly speaking, the Dvoretzky–Rogers Lemma builds something like a cube around $K$, at least in a subspace of dimension about $\frac{n}{2}$, to which we then apply the result for the cube. However, I cannot resist mentioning that the methods of Lecture 6, involving sharp convolution inequalities, can be used to show that among all symmetric bodies $K$ with maximal ellipsoid $B_2^n$ the cube is precisely the one for which the integral in (9.1) is smallest. This is proved by showing that for each $r$, the Gaussian measure of $rK$ is at most that of $[-r, r]^n$. The details are left to the reader.

This last lecture has described work that dates back to the seventies. Although some of the material in earlier lectures is more recent (and some is much older), I have really only scratched the surface of what has been done in the last twenty years. The book of Pisier to which I have referred several times gives a more comprehensive account of many of the developments. I hope that readers of these notes may feel motivated to discover more.

## Acknowledgements

I would like to thank Silvio Levy for his help in the preparation of these notes, and one of the workshop participants, John Mount, for proofreading the notes and suggesting several improvements. Finally, a very big thank you to my wife Sachiko Kusukawa for her great patience and constant love.

## References

[Ball 1990] K. M. Ball, "Volumes of sections of cubes and related problems", pp. 251–260 in *Geometric aspects of functional analysis* (Israel Seminar, 1987–1988), edited by J. Lindenstrauss and V. D. Milman, Lecture Notes in Math. **1376**, Springer, 1990.

[Ball 1991] K. M. Ball, "Volume ratios and a reverse isoperimetric inequality", *J. London Math. Soc.* **44** (1991), 351–359.

[Beckner 1975] W. Beckner, "Inequalities in Fourier analysis", *Ann. of Math.* **102** (1975), 159–182.

[Borell 1975] C. Borell, "The Brunn–Minkowski inequality in Gauss space", *Inventiones Math.* **30** (1975), 205–216.

[Bourgain and Milman 1987] J. Bourgain and V. Milman, "New volume ratio properties for convex symmetric bodies in $\mathbb{R}^n$", *Invent. Math.* **88** (1987), 319–340.

[Bourgain et al. 1989] J. Bourgain, J. Lindenstrauss, and V. Milman, "Approximation of zonoids by zonotopes", *Acta Math.* **162** (1989), 73–141.

[Brascamp and Lieb 1976a] H. J. Brascamp and E. H. Lieb, "Best constants in Young's inequality, its converse and its generalization to more than three functions", *Advances in Math.* **20** (1976), 151–173.

[Brascamp and Lieb 1976b] H. J. Brascamp and E. H. Lieb, "On extensions of the Brunn–Minkowski and Prékopa–Leindler theorems, including inequalities for log concave functions, and with an application to the diffusion equation", *J. Funct. Anal.* **22** (1976), 366–389.

[Brøndsted 1983] A. Brøndsted, *An introduction to convex polytopes*, Graduate Texts in Math. **90**, Springer, New York, 1983.

[Busemann 1949]  H. Busemann, "A theorem on convex bodies of the Brunn–Minkowski type", *Proc. Nat. Acad. Sci. USA* **35** (1949), 27–31.

[Figiel et al. 1977]  T. Figiel, J. Lindenstrauss, and V. Milman, "The dimension of almost spherical sections of convex bodies", *Acta Math.* **139** (1977), 53–94.

[Garnaev and Gluskin 1984]  A. Garnaev and E. Gluskin, "The widths of a Euclidean ball", *Dokl. A. N. USSR* **277** (1984), 1048–1052. In Russian.

[Gordon 1985]  Y. Gordon, "Some inequalities for Gaussian processes and applications", *Israel J. Math.* **50** (1985), 265–289.

[Hoeffding 1963]  W. Hoeffding, "Probability inequalities for sums of bounded random variables", *J. Amer. Statist. Assoc.* **58** (1963), 13–30.

[John 1948]  F. John, "Extremum problems with inequalities as subsidiary conditions", pp. 187–204 in *Studies and essays presented to R. Courant on his 60th birthday* (Jan. 8, 1948), Interscience, New York, 1948.

[Johnson and Schechtman 1982]  W. B. Johnson and G. Schechtman, "Embedding $\ell_p^m$ into $\ell_1^n$", *Acta Math.* **149** (1982), 71–85.

[Kašin 1977]  B. S. Kašin, "The widths of certain finite-dimensional sets and classes of smooth functions", *Izv. Akad. Nauk SSSR Ser. Mat.* **41**:2 (1977), 334–351, 478. In Russian.

[Lieb 1990]  E. H. Lieb, "Gaussian kernels have only Gaussian maximizers", *Invent. Math.* **102** (1990), 179–208.

[Liûsternik 1935]  L. A. Liûsternik, "Die Brunn–Minkowskische Ungleichung für beliebige messbare Mengen", *C. R. Acad. Sci. URSS* **8** (1935), 55–58.

[Maurey 1991]  B. Maurey, "Some deviation inequalities", *Geom. Funct. Anal.* **1**:2 (1991), 188–197.

[Milman 1971]  V. Milman, "A new proof of A. Dvoretzky's theorem on cross-sections of convex bodies", *Funkcional. Anal. i Priložen* **5** (1971), 28–37. In Russian.

[Milman 1985]  V. Milman, "Almost Euclidean quotient spaces of subspaces of finite dimensional normed spaces", *Proc. Amer. Math. Soc.* **94** (1985), 445–449.

[Pisier 1989]  G. Pisier, *The volume of convex bodies and Banach space geometry*, Tracts in Math. **94**, Cambridge U. Press, Cambridge, 1989.

[Rogers 1964]  C. A. Rogers, *Packing and covering*, Cambridge U. Press, Cambridge, 1964.

[Schneider 1993]  R. Schneider, *Convex bodies: the Brunn–Minkowski theory*, Encyclopedia of Math. and its Applications **44**, Cambridge U. Press, 1993.

[Szarek 1978]  S. J. Szarek, "On Kashin's almost Euclidean orthogonal decomposition of $\ell_1^n$", *Bull. Acad. Polon. Sci. Sér. Sci. Math. Astronom. Phys.* **26** (1978), 691–694.

[Talagrand 1988]  M. Talagrand, "An isoperimetric inequality on the cube and the Khintchine-Kahane inequalities", *Proc. Amer. Math. Soc.* **104** (1988), 905–909.

[Talagrand 1990]  M. Talagrand, "Embedding subspaces of $L_1$ into $\ell_1^N$", *Proc. Amer. Math. Soc.* **108** (1990), 363–369.

[Talagrand 1991a]  M. Talagrand, "A new isoperimetric inequality", *Geom. Funct. Anal.* **1**:2 (1991), 211–223.

[Talagrand 1991b]  M. Talagrand, "A new isoperimetric inequality and the concentration of measure phenomenon", pp. 94–124 in *Geometric aspects of functional analysis* (Israel Seminar, 1989–1990), edited by J. Lindenstrauss and V. D. Milman, Lecture Notes in Math. **1469**, Springer, 1991.

[Tomczak-Jaegermann 1988]  N. Tomczak-Jaegermann, *Banach–Mazur distances and finite-dimensional operator ideals*, Pitman Monographs and Surveys in Pure and Applied Math. **38**, Longman, Harlow, 1988.

[Williams 1991]  D. Williams, *Probability with martingales*, Cambridge Mathematical Textbooks, Cambridge University Press, Cambridge, 1991.

# Index

KEITH BALL
DEPARTMENT OF MATHEMATICS
UNIVERSITY COLLEGE
UNIVERSITY OF LONDON
LONDON
UNITED KINGDOM
  kmb@math.ucl.ac.uk

Flavors of Geometry
MSRI Publications
Volume **31**, 1997

# Hyperbolic Geometry

JAMES W. CANNON, WILLIAM J. FLOYD, RICHARD KENYON,
AND WALTER R. PARRY

## CONTENTS

## 1. Introduction

Hyperbolic geometry was created in the first half of the nineteenth century
in the midst of attempts to understand Euclid's axiomatic basis for geometry.
It is one type of *non-Euclidean geometry*, that is, a geometry that discards one
of Euclid's axioms. Einstein and Minkowski found in non-Euclidean geometry a

This work was supported in part by The Geometry Center, University of Minnesota, an STC
funded by NSF, DOE, and Minnesota Technology, Inc., by the Mathematical Sciences Research
Institute, and by NSF research grants.

geometric basis for the understanding of physical time and space. In the early part of the twentieth century every serious student of mathematics and physics studied non-Euclidean geometry. This has not been true of the mathematicians and physicists of our generation. Nevertheless with the passage of time it has become more and more apparent that the negatively curved geometries, of which hyperbolic non-Euclidean geometry is the prototype, are the generic forms of geometry. They have profound applications to the study of complex variables, to the topology of two- and three-dimensional manifolds, to the study of finitely presented infinite groups, to physics, and to other disparate fields of mathematics. A working knowledge of hyperbolic geometry has become a prerequisite for workers in these fields.

These notes are intended as a relatively quick introduction to hyperbolic geometry. They review the wonderful history of non-Euclidean geometry. They give five different analytic models for and several combinatorial approximations to non-Euclidean geometry by means of which the reader can develop an intuition for the behavior of this geometry. They develop a number of the properties of this geometry that are particularly important in topology and group theory. They indicate some of the fundamental problems being approached by means of non-Euclidean geometry in topology and group theory.

Volumes have been written on non-Euclidean geometry, which the reader must consult for more exhaustive information. We recommend [Iversen 1993] for starters, and [Benedetti and Petronio 1992; Thurston 1997; Ratcliffe 1994] for more advanced readers. The latter has a particularly comprehensive bibliography.

## 2. The Origins of Hyperbolic Geometry

Except for Euclid's five fundamental postulates of plane geometry, which we paraphrase from [Kline 1972], most of the following historical material is taken from Felix Klein's book [1928]. Here are Euclid's postulates in contemporary language (compare [Euclid 1926]):

1. Each pair of points can be joined by one and only one straight line segment.
2. Any straight line segment can be indefinitely extended in either direction.
3. There is exactly one circle of any given radius with any given center.
4. All right angles are congruent to one another.
5. If a straight line falling on two straight lines makes the interior angles on the same side less than two right angles, the two straight lines, if extended indefinitely, meet on that side on which the angles are less than two right angles.

Of these five postulates, the fifth is by far the most complicated and unnatural. Given the first four, the fifth postulate can easily be seen to be equivalent to the following *parallel postulate*, which explains why the expressions "Euclid's fifth postulate" and "the parallel parallel" are often used interchangeably:

5′. Given a line and a point not on it, there is exactly one line going through the given point that is parallel to the given line.

For two thousand years mathematicians attempted to deduce the fifth postulate from the four simpler postulates. In each case one reduced the proof of the fifth postulate to the conjunction of the first four postulates with an additional natural postulate that, in fact, proved to be equivalent to the fifth:

Proclus (ca. 400 A.D.) used as additional postulate the assumption that the points at constant distance from a given line on one side form a straight line.

The Englishman John Wallis (1616–1703) used the assumption that to every triangle there is a similar triangle of each given size.

The Italian Girolamo Saccheri (1667–1733) considered quadrilaterals with two base angles equal to a right angle and with vertical sides having equal length and deduced consequences from the (non-Euclidean) possibility that the remaining two angles were not right angles.

Johann Heinrich Lambert (1728–1777) proceeded in a similar fashion and wrote an extensive work on the subject, posthumously published in 1786.

Göttingen mathematician Kästner (1719–1800) directed a thesis of student Klügel (1739–1812), which considered approximately thirty proof attempts for the parallel postulate.

Decisive progress came in the nineteenth century, when mathematicians abandoned the effort to find a contradiction in the denial of the fifth postulate and instead worked out carefully and completely the consequences of such a denial. It was found that a coherent theory arises if instead one assumes that

Given a line and a point not on it, there is more than one line going through the given point that is parallel to the given line.

This postulate is to hyperbolic geometry as the parallel postulate 5′ is to Euclidean geometry.

Unusual consequences of this change came to be recognized as fundamental and surprising properties of non-Euclidean geometry: equidistant curves on either side of a straight line were in fact not straight but curved; similar triangles were congruent; angle sums in a triangle were not equal to $\pi$, and so forth.

That the parallel postulate fails in the models of non-Euclidean geometry that we shall give will be apparent to the reader. The unusual properties of non-Euclidean geometry that we have mentioned will all be worked out in Section 13, entitled "Curious facts about hyperbolic space".

History has associated five names with this enterprise, those of three professional mathematicians and two amateurs.

The amateurs were jurist Schweikart and his nephew Taurinus (1794–1874). By 1816 Schweikart had developed, in his spare time, an "astral geometry" that was independent of the fifth postulate. His nephew Taurinus had attained a non-Euclidean hyperbolic geometry by the year 1824.

The professionals were Carl Friedrich Gauss (1777–1855), Nikolaĭ Ivanovich Lobachevskiĭ (1793–1856), and János (or Johann) Bolyai (1802–1860). From the papers of his estate it is apparent that Gauss had considered the parallel postulate extensively during his youth and at least by the year 1817 had a clear picture of non-Euclidean geometry. The only indications he gave of his knowledge were small comments in his correspondence. Having satisfied his own curiosity, he was not interested in defending the concept in the controversy that was sure to accompany its announcement. Bolyai's father Fárkás (or Wolfgang) (1775–1856) was a student friend of Gauss and remained in correspondence with him throughout his life. Fárkás devoted much of his life's effort unsuccessfully to the proof of the parallel postulate and consequently tried to turn his son away from its study. Nevertheless, János attacked the problem with vigor and had constructed the foundations of hyperbolic geometry by the year 1823. His work appeared in 1832 or 1833 as an appendix to a textbook written by his father. Lobachevskiĭ also developed a non-Euclidean geometry extensively and was, in fact, the first to publish his findings, in 1829. See [Lobachevskiĭ 1898; Bolyai and Bolyai 1913].

Gauss, the Bolyais, and Lobachevskiĭ developed non-Euclidean geometry axiomatically on a synthetic basis. They had neither an analytic understanding nor an analytic model of non-Euclidean geometry. They did not prove the *consistency* of their geometries. They instead satisfied themselves with the conviction they attained by extensive exploration in non-Euclidean geometry where theorem after theorem fit consistently with what they had discovered to date. Lobachevskiĭ developed a non-Euclidean trigonometry that paralleled the trigonometric formulas of Euclidean geometry. He argued for the consistency based on the consistency of his analytic formulas.

The basis necessary for an analytic study of hyperbolic non-Euclidean geometry was laid by Leonhard Euler, Gaspard Monge, and Gauss in their studies of curved surfaces. In 1837 Lobachevskiĭ suggested that curved surfaces of constant negative curvature might represent non-Euclidean geometry. Two years later, working independently and largely in ignorance of Lobachevskiĭ's work, yet publishing in the same journal, Minding made an extensive study of surfaces of constant curvature and verified Lobachevskiĭ suggestion. Bernhard Riemann (1826–1866), in his vast generalization [Riemann 1854] of curved surfaces to the study of what are now called Riemannian manifolds, recognized all of these relationships and, in fact, to some extent used them as a springboard for his studies. All of the connections among these subjects were particularly pointed out by Eugenio Beltrami in 1868. This analytic work provided specific analytic models for non-Euclidean geometry and established the fact that non-Euclidean geometry was precisely as consistent as Euclidean geometry itself.

We shall consider in this exposition five of the most famous of the analytic models of hyperbolic geometry. Three are conformal models associated with the name of Henri Poincaré. A conformal model is one for which the metric is a point-by-point scaling of the Euclidean metric. Poincaré discovered his models in the process of defining and understanding Fuchsian, Kleinian, and general automorphic functions of a single complex variable. The story is one of the most famous and fascinating stories about discovery and the work of the subconscious mind in all of science. We quote from [Poincaré 1908]:

> For fifteen days I strove to prove that there could not be any functions like those I have since called Fuchsian functions. I was then very ignorant; every day I seated myself at my work table, stayed an hour or two, tried a great number of combinations and reached no results. One evening, contrary to my custom, I drank black coffee and could not sleep. Ideas rose in crowds; I felt them collide until pairs interlocked, so to speak, making a stable combination. By the next morning I had established the existence of a class of Fuchsian functions, those which come from the hypergeometric series; I had only to write out the results, which took but a few hours.
>
> Then I wanted to represent these functions by the quotient of two series; this idea was perfectly conscious and deliberate, the analogy with elliptic functions guided me. I asked myself what properties these series must have if they existed, and I succeeded without difficulty in forming the series I have called theta-Fuchsian.
>
> Just at this time I left Caen, where I was then living, to go on a geological excursion under the auspices of the school of mines. The changes of travel made me forget my mathematical work. Having reached Coutances, we entered an omnibus to go some place or other. At the moment when I put my foot on the step the idea came to me, without anything in my former thoughts seeming to have paved the way for it, that the transformations I had used to define the Fuchsian functions were identical with those of non-Euclidean geometry. I did not verify the idea; I should not have had time, as, upon taking my seat in the omnibus, I went on with a conversation already commenced, but I felt a perfect certainty. On my return to Caen, for conscience' sake I verified the result at my leisure.

## 3. Why Call it Hyperbolic Geometry?

The non-Euclidean geometry of Gauss, Lobachevskiĭ, and Bolyai is usually called *hyperbolic geometry* because of one of its very natural analytic models. We describe that model here.

Classically, space and time were considered as independent quantities; an event could be given coordinates $(x_1, \ldots, x_{n+1}) \in \mathbb{R}^{n+1}$, with the coordinate $x_{n+1}$ representing time, and the only reasonable metric was the Euclidean metric with the positive definite square-norm $x_1^2 + \cdots + x_{n+1}^2$.

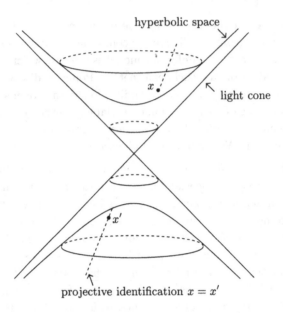

**Figure 1.** Minkowski space.

Relativity changed all that; in flat space-time geometry the speed of light should be constant as viewed from any inertial reference frame. The Minkowski model for space-time geometry is again $\mathbb{R}^{n+1}$ but with the indefinite norm $x_1^2 + \cdots + x_n^2 - x_{n+1}^2$ defining distance. The *light cone* is defined as the set of points of norm 0. For points $(x_1, \ldots, x_n, x_{n+1})$ on the light cone, the Euclidean space-distance

$$(x_1^2 + \cdots + x_n^2)^{1/2}$$

from the origin is equal to the time $x_{n+1}$ from the origin; this equality expresses the constant speed of light starting at the origin.

These norms have associated inner products, denoted $\cdot$ for the Euclidean inner product and $*$ for the non-Euclidean.

If we consider the set of points at constant squared distance from the origin, we obtain in the Euclidean case the spheres of various radii and in Minkowski space hyperboloids of one or two sheets. We may thus define the unit *n-dimensional sphere* in Euclidean space $\mathbb{R}^{n+1}$ by the formula $S^n = \{x \in \mathbb{R}^{n+1} : x \cdot x = 1\}$ and *n-dimensional hyperbolic space* by the formula $\{x \in \mathbb{R}^{n+1} : x * x = -1\}$. Thus hyperbolic space is a hyperboloid of two sheets that may be thought of as a "sphere" of squared radius $-1$ or of radius $i = \sqrt{-1}$; hence the name hyperbolic geometry. See Figure 1.

Usually we deal only with one of the two sheets of the hyperboloid or identify the two sheets projectively.

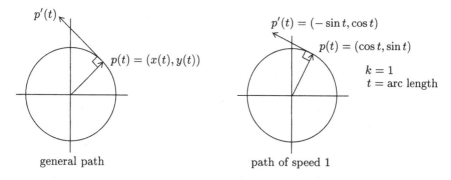

**Figure 2.** The circle $S^1$.

## 4. Understanding the One-Dimensional Case

The key to understanding hyperbolic space $H^n$ and its intrinsic metric coming from the indefinite Minkowski inner product $*$ is to first understand the case $n = 1$. We argue by analogy with the Euclidean case and prepare the analogy by recalling the familiar Euclidean case of the circle $S^1$.

Let $p : (-\infty, \infty) \to S^1$ be a smooth path with $p(0) = (1, 0)$. If we write in coordinates $p(t) = (x(t), y(t))$ where $x^2 + y^2 = 1$, then differentiating this equation we find

$$2x(t)x'(t) + 2y(t)y'(t) = 0,$$

or in other words $p(t) \cdot p'(t) = 0$. That is, the velocity vector $p'(t)$ is Euclidean-perpendicular to the position vector $p(t)$. In particular we may write $p'(t) = k(t)(-y(t), x(t))$, since the tangent space to $S^1$ at $p(t)$ is one-dimensional and $(-y(t), x(t))$ is Euclidean-perpendicular to $p = (x, y)$. See Figure 2.

If we assume in addition that $p(t)$ has *constant* speed 1, then

$$1 = |p'(t)| = |k(t)|\sqrt{(-y)^2 + x^2} = |k(t)|,$$

and so $k \equiv \pm 1$. Taking $k \equiv 1$, we see that $p = (x, y)$ travels around the unit circle in the Euclidean plane at constant speed 1. Consequently we may by definition identify $t$ with Euclidean arclength on the unit circle, $x = x(t)$ with $\cos t$ and $y = y(t)$ with $\sin t$, and we see that we have given a complete proof of the fact from beginning calculus that the derivative of the cosine is minus the sine and that the derivative of the sine is the cosine, a proof that is conceptually simpler than the proofs usually given in class.

In formulas, taking $k = 1$, we have shown that $x$ and $y$ (the cosine and sine) satisfy the system of differential equations

$$x'(t) = -y(t), \qquad y'(t) = x(t),$$

with initial conditions $x(0) = 1$, $y(0) = 0$. We then need only apply some elementary method such as the method of undetermined coefficients to easily discover the classical power series for the sine and cosine:

$$\cos t = 1 - t^2/2! + t^4/4! - \cdots,$$
$$\sin t = t - t^3/3! + t^5/5! - \cdots.$$

The hyperbolic calculation in $H^1$ requires only a new starting point $(0,1)$ instead of $(1,0)$, the replacement of $S^1$ by $H^1$, the replacement of the Euclidean inner product $\cdot$ by the hyperbolic inner product $*$, an occasional replacement of $+1$ by $-1$, the replacement of Euclidean arclength by hyperbolic arclength, the replacement of cosine by hyperbolic sine, and the replacement of sine by the hyperbolic cosine. Here is the calculation.

Let $p : (-\infty, \infty) \to H^1$ be a smooth path with $p(0) = (0,1)$. If we write in coordinates $p(t) = (x(t), y(t))$ where $x^2 - y^2 = -1$, then differentiating this equation we find

$$2x(t)x'(t) - 2y(t)y'(t) = 0;$$

in other words $p(t) * p'(t) = 0$. That is, the velocity vector $p'(t)$ is hyperbolic-perpendicular to the position vector $p(t)$. In particular we may write $p'(t) = k(t)(y(t), x(t))$, since the tangent space to $H^1$ at $p(t)$ is one-dimensional and the vector $(y(t), x(t))$ is hyperbolic-perpendicular to $p = (x, y)$. See Figure 3.

If we assume in addition that $p(t)$ has *constant speed* 1, then $1 = |p'(t)| = |k(t)|\sqrt{y^2 - x^2} = |k(t)|$, and so $k \equiv \pm 1$. Taking $k \equiv 1$, we see that $p = (x, y)$ travels to the right along the "unit" hyperbola in the Minkowski plane at constant hyperbolic speed 1. Consequently we may by definition identify $t$ with hyperbolic arclength on the unit hyperbola $H^1$, $x = x(t)$ with $\sinh t$ and $y = y(t)$ with $\cosh t$, and we see that we have given a complete proof of the fact from beginning calculus that the derivative of the hyperbolic cosine is the hyperbolic sine and that the derivative of the hyperbolic sine is the hyperbolic cosine, a proof that is conceptually simpler than the proofs usually given in class.

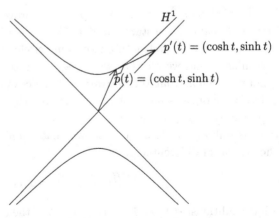

**Figure 3.** The hyperbolic line $H^1$.

In formulas, taking $k = 1$, we have shown that $x$ and $y$ (the hyperbolic sine and cosine) satisfy the system of differential equations

$$x'(t) = y(t), \qquad y'(t) = x(t),$$

with initial conditions $x(0) = 0$, $y(0) = 1$. We then need only apply some elementary method such as the method of undetermined coefficients to easily discover the classical power series for the hyperbolic sine and cosine:

$$\cosh t = 1 + t^2/2! + t^4/4! + \cdots,$$
$$\sinh t = t + t^3/3! + t^5/5! + \cdots.$$

It seems to us a shame that these analogies, being as easy as they are, are seldom developed in calculus classes. The reason of course is that the analogies become forced if one is not willing to leave the familiar Euclidean plane for the unfamiliar Minkowski plane.

Note the remarkable fact that our calculation showed that a nonzero tangent vector to $H^1$ has *positive square norm* with respect to the indefinite inner product $*$; that is, the indefinite inner product on the Minkowski plane restricts to a positive definite inner product on hyperbolic one-space. We shall find that the analogous result is true in higher dimensions and that the formulas we have calculated for hyperbolic length in dimension one apply in the higher-dimensional setting as well.

## 5. Generalizing to Higher Dimensions

In higher dimensions, $H^n$ sits inside $\mathbb{R}^{n+1}$ as a hyperboloid. If $p : (-\infty, \infty) \to H^n$ again describes a smooth path, then from the defining equations we still have $p(t) * p'(t) = 0$. By taking paths in any direction running through the point $p(t)$, we see that the tangent vectors to $H^n$ at $p(t)$ form the hyperbolic orthogonal complement to the vector $p(t)$ (vectors are hyperbolically orthogonal if their inner product with respect to $*$ is 0).

We can show that the form $*$ restricted to the tangent space is positive definite in either of two instructive ways.

The first method uses the Cauchy–Schwarz inequality $(x \cdot y)^2 \leq (x \cdot x)(y \cdot y)$. Suppose that $p = (\hat{p}, p_{n+1})$ is in $H^n$ and $x = (\hat{x}, x_{n+1}) \neq 0$ is in the tangent space of $H^n$ at $p$, where $\hat{p}, \hat{x} \in \mathbb{R}^n$. If $x_{n+1} = 0$, then $x * x = x \cdot x$. Hence $x * x > 0$ if $x_{n+1} = 0$, so we may assume that $x_{n+1} \neq 0$. Then $0 = x * p = \hat{x} \cdot \hat{p} - x_{n+1} p_{n+1}$, and $-1 = p * p = \hat{p} \cdot \hat{p} - p_{n+1}^2$. Hence, Cauchy–Schwarz gives

$$(\hat{x} \cdot \hat{x})(\hat{p} \cdot \hat{p}) \geq (\hat{x} \cdot \hat{p})^2 = (x_{n+1} p_{n+1})^2 = x_{n+1}^2 (\hat{p} \cdot \hat{p} + 1).$$

Therefore, $(x * x)(\hat{p} \cdot \hat{p}) \geq x_{n+1}^2$, which implies $x * x > 0$ if $x \neq 0$.

The second method analyzes the inner product $*$ algebraically. (For complete details, see [Weyl 1919], for example.) Take a basis $p, p_1, \ldots, p_n$ for $\mathbb{R}^{n+1}$ where $p$ is the point of interest in $H^n$ and the remaining vectors span the $n$-dimensional tangent space to $H^n$ at $p$. Now apply the Gram–Schmidt orthogonalization process to this basis. Since $p*p = -1$ by the defining equation for $H^n$, the vector $p$, being already a unit vector, is unchanged by the process and the remainder of the resulting basis spans the orthogonal complement of $p$, which is the tangent space to $H^n$ at $p$. Since the inner product $*$ is nondegenerate, the resulting matrix is diagonal with entries of $\pm 1$ on the diagonal, one of the $-1$'s corresponding to the vector $p$. By Sylvester's theorem of inertia, the number of $+1$'s and $-1$'s on the diagonal is an invariant of the inner product (the number of $1$'s is the dimension of the largest subspace on which the metric is positive definite). But with the standard basis for $\mathbb{R}^{n+1}$, there is exactly one $-1$ on the diagonal and the remaining entries are $+1$. Hence the same is true of our basis. Thus the matrix of the inner product when restricted to our tangent space is the identity matrix of order $n$; that is, the restriction of the metric to the tangent space is positive definite.

Thus the inner product $*$ restricted to $H^n$ defines a genuine Riemannian metric on $H^n$.

## 6. Rudiments of Riemannian Geometry

Our analytic models of hyperbolic geometry will all be differentiable manifolds with a Riemannian metric.

One first defines a Riemannian metric and associated geometric notions on Euclidean space. A Riemannian metric $ds^2$ on Euclidean space $\mathbb{R}^n$ is a function that assigns at each point $p \in \mathbb{R}^n$ a positive definite symmetric inner product on the tangent space at $p$, this inner product varying differentiably with the point $p$. Given this inner product, it is possible to define any number of standard geometric notions such as the length $|x|$ of a vector $x$, where $|x|^2 = x \cdot x$, the angle $\theta$ between two vectors $x$ and $y$, where $\cos\theta = (x\cdot y)/(|x|\cdot|y|)$, the length element $ds = \sqrt{ds^2}$, and the area element $dA$, where $dA$ is calculated as follows: if $x_1, \ldots, x_n$ are the standard coordinates on $\mathbb{R}^n$, then $ds^2$ has the form $\sum_{i,j} g_{ij}\, dx_i\, dx_j$, and the matrix $(g_{ij})$ depends differentiably on $x$ and is positive definite and symmetric. Let $\sqrt{|g|}$ denote the square root of the determinant of $(g_{ij})$. Then $dA = \sqrt{|g|}\, dx_1\, dx_2 \cdots dx_n$. If $f : \mathbb{R}^k \to \mathbb{R}^n$ is a differentiable map, one can define the pullback $f^*(ds^2)$ by the formula

$$f^*(ds^2)(v, w) = ds^2(Df(v), Df(w))$$

where $v$ and $w$ are tangent vectors at a point $u$ of $\mathbb{R}^k$ and $Df$ is the derivative map that takes tangent vectors at $u$ to tangent vectors at $x = f(u)$. One can also calculate the pullback formally by replacing $g_{ij}(x)$ with $x \in \mathbb{R}^n$ by $g_{ij}\circ f(u)$, where $u \in \mathbb{R}^k$ and $f(u) = x$, and replacing $dx_i$ by $\sum_j(\partial f_i/\partial u_j)du_j$. One can

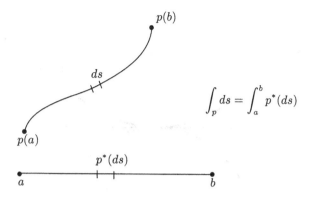

$$\int_p ds = \int_a^b p^*(ds)$$

**Figure 4.** The length of a path.

calculate the length of a path $p : [a, b] \to \mathbb{R}^n$ by integrating $ds$ over $p$:

$$\int_p ds = \int_a^b p^*(ds).$$

See Figure 4. The Riemannian distance $d(p, q)$ between two points $p$ and $q$ in $\mathbb{R}^n$ is defined as the infimum of path length over all paths joining $p$ and $q$.

Finally, one generalizes all of these notions to manifolds by requiring the existence of a Riemannian metric on each coordinate chart with these metrics being invariant under pullback on transition functions connecting these charts; that is, if $ds_1^2$ is the Riemannian metric on chart one and if $ds_2^2$ is the Riemannian metric on chart two and if $f$ is a transition function connecting these two charts, then $f^*(ds_2^2) = ds_1^2$. The standard change of variables formulas from calculus show that path lengths and areas are invariant under chart change.

## 7. Five Models of Hyperbolic Space

We describe here five analytic models of hyperbolic space. The theory of hyperbolic geometry could be built in a unified way within a single model, but with several models it is as if one were able to turn the object that is hyperbolic space about in one's hands so as to see it first from above, then from the side, and finally from beneath or within; each view supplies its own natural intuitions. Each model has its own metric, geodesics, isometries, and so on. Here are our mnemonic names for the five models:

$H$, the Half-space model.

$I$, the Interior of the disk model.

$J$, the Jemisphere model (pronounce the J as in Spanish).

$K$, the Klein model.

$L$, the 'Loid model (short for hyperboloid).

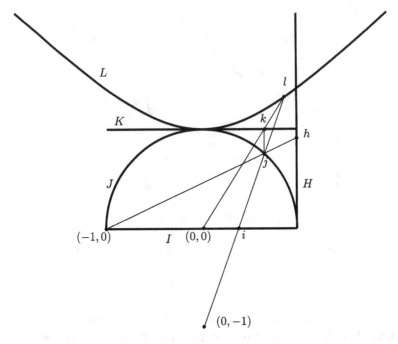

**Figure 5.** The five analytic models and their connecting isometries. The points $h \in H$, $i \in I$, $j \in J$, $k \in K$, and $l \in L$ can be thought of as the same point in (synthetic) hyperbolic space.

Each model is defined on a different subset of $\mathbb{R}^{n+1}$, called its *domain;* for $n = 1$ these sets are schematically indicated in Figure 5, which can also be regarded as a cross section of the picture in higher dimensions. Here are the definitions of the five domains:

$$H = \{(1, x_2, \ldots, x_{n+1}) : x_{n+1} > 0\};$$
$$I = \{(x_1, \ldots, x_n, 0) : x_1^2 + \cdots + x_n^2 < 1\};$$
$$J = \{(x_1, \ldots, x_{n+1}) : x_1^2 + \cdots + x_{n+1}^2 = 1 \text{ and } x_{n+1} > 0\};$$
$$K = \{(x_1, \ldots, x_n, 1) : x_1^2 + \cdots + x_n^2 < 1\};$$
$$L = \{(x_1, \ldots, x_n, x_{n+1}) : x_1^2 + \cdots + x_n^2 - x_{n+1}^2 = -1 \text{ and } x_{n+1} > 0\}.$$

The associated Riemannian metrics $ds^2$ that complete the analytic description of the five models are:

$$ds_H^2 = \frac{dx_2^2 + \cdots + dx_{n+1}^2}{x_{n+1}^2};$$

$$ds_I^2 = 4\frac{dx_1^2 + \cdots + dx_n^2}{(1 - x_1^2 - \cdots - x_n^2)^2};$$

$$ds_J^2 = \frac{dx_1^2 + \cdots + dx_{n+1}^2}{x_{n+1}^2};$$

$$ds_K^2 = \frac{dx_1^2 + \cdots + dx_n^2}{(1 - x_1^2 - \cdots - x_n^2)} + \frac{(x_1\, dx_1 + \cdots + x_n\, dx_n)^2}{(1 - x_1^2 - \cdots - x_n^2)^2};$$

$$ds_L^2 = dx_1^2 + \cdots + dx_n^2 - dx_{n+1}^2.$$

To see that these five models are isometrically equivalent, we need to describe isometries among them. We use $J$ as the central model and describe for each of the others a simple map to or from $J$.

The map $\alpha : J \to H$ is central projection from the point $(-1, 0, \ldots, 0)$:

$$\alpha : J \to H, \quad (x_1, \ldots, x_{n+1}) \mapsto (1, 2x_2/(x_1 + 1), \ldots, 2x_{n+1}/(x_1 + 1)).$$

The map $\beta : J \to I$ is central projection from $(0, \ldots, 0, -1)$:

$$\beta : J \to I, \quad (x_1, \ldots, x_{n+1}) \mapsto (x_1/(x_{n+1} + 1), \ldots, x_n/(x_{n+1} + 1), 0).$$

The map $\gamma : K \to J$ is vertical projection:

$$\gamma : K \to J, \quad (x_1, \ldots, x_n, 1) \mapsto \left(x_1, \ldots, x_n, \sqrt{1 - x_1^2 - \cdots - x_n^2}\,\right).$$

The map $\delta : L \to J$ is central projection from $(0, \ldots, 0, -1)$:

$$\delta : L \to J, \quad (x_1, \ldots, x_{n+1}) \mapsto (x_1/x_{n+1}, \ldots, x_n/x_{n+1}, 1/x_{n+1}).$$

Each map can be used in the standard way to pull back the Riemannian metric from the target domain to the source domain and to verify thereby that the maps are isometries. Among the twenty possible connecting maps among our models, we have chosen the four for which we find the calculation of the metric pullback easiest. It is worth noting that the metric on the Klein model $K$, which has always struck us as particularly ugly and unintuitive, takes on obvious meaning and structure relative to the metric on $J$ from which it naturally derives via the connecting map $\gamma : K \to J$. We perform here two of the four pullback calculations as examples and recommend that the reader undertake the other two.

Here is the calculation that shows that $\alpha^*(ds_H^2) = ds_J^2$. Set

$$y_2 = 2x_2/(x_1 + 1), \quad \ldots, \quad y_{n+1} = 2x_{n+1}/(x_1 + 1).$$

Then

$$dy_i = \frac{2}{x_1 + 1}\left(dx_i - \frac{x_i}{x_1 + 1}\, dx_1\right).$$

Since $x_1^2 + \cdots + x_{n+1}^2 = 1$,

$$x_1 \, dx_1 = -(x_2 \, dx_2 + \cdots + x_{n+1} \, dx_{n+1})$$

and

$$x_2^2 + \cdots + x_{n+1}^2 = 1 - x_1^2.$$

These equalities justify the following simple calculation:

$$\alpha^*(ds_H^2) = \frac{1}{y_{n+1}^2}(dy_2^2 + \cdots + dy_{n+1}^2)$$

$$= \frac{(x_1+1)^2}{4x_{n+1}^2} \cdot \frac{4}{(x_1+1)^2}\left(\sum_{i=2}^{n+1} dx_i^2 - \frac{2dx_1}{x_1+1}\sum_{i=2}^{n+1} x_i \, dx_i + \frac{dx_1^2}{(x_1+1)^2}\sum_{i=2}^{n+1} x_i^2\right)$$

$$= \frac{1}{x_{n+1}^2}\left(\sum_{i=2}^{n+1} dx_i^2 + \frac{2}{(x_1+1)} \cdot x_1 \, dx_1^2 + \frac{dx_1^2}{(x_1+1)^2}(1-x_1^2)\right)$$

$$= \frac{1}{x_{n+1}^2}\sum_{i=1}^{n+1} dx_i^2$$

$$= ds_J^2.$$

Here is the calculation that shows that $\gamma^*(ds_J^2) = ds_K^2$. Set $y_1 = x_1, \ldots,$ $y_n = x_n$, and $y_{n+1}^2 = 1 - y_1^2 - \cdots - y_n^2 = 1 - x_1^2 - \cdots - x_n^2$. Then $dy_i = dx_i$ for $i = 1, \ldots, n$ and $y_{n+1} \, dy_{n+1} = -(x_1 \, dx_1 + \cdots + x_n \, dx_n)$. Thus

$$\gamma^*(ds_J^2) = \frac{1}{y_{n+1}^2}(dy_1^2 + \cdots + dy_n^2) + \frac{1}{y_{n+1}^2}dy_{n+1}^2$$

$$= \frac{1}{(1 - x_1^2 - \cdots - x_n^2)}(dx_1^2 + \cdots + dx_n^2) + \frac{(x_1 \, dx_1 + \cdots + x_n \, dx_n)^2}{(1 - x_1^2 - \cdots - x_n^2)^2}$$

$$= ds_K^2.$$

The other two pullback computations are comparable.

## 8. Stereographic Projection

In order to understand the relationships among these models, it is helpful to understand the geometric properties of the connecting maps. Two of them are *central* or *stereographic* projection from a sphere to a plane. In this section we develop some important properties of stereographic projection. We begin with the definition and then establish the important properties that stereographic projection *preserves angles* and *takes spheres to planes or spheres*. We give a geometric proof in dimension three and an analytic proof in general.

DEFINITION. Let $S^n$ denote a sphere of dimension $n$ in Euclidean $(n + 1)$-dimensional space $\mathbb{R}^{n+1}$. Let $P$ denote a plane tangent to the sphere $S^n$ at the point $S$, which we think of as the south pole of $S^n$. Let $N$ denote the point of $S^n$ opposite $S$, a point that we think of as the north pole of $S^n$. If $x$ is any point of $S^n \setminus \{N\}$, there is a unique point $\pi(x)$ of $P$ on the line that contains $N$ and $x$. It is called the *stereographic projection* from $x$ into $P$. See Figure 6. Note that $\pi$ has a natural extension, also denoted by $\pi$, which takes all of $\mathbb{R}^{n+1}$ except for the plane $\{x : x_{n+1} = 1\}$ into $P$.

THEOREM 8.1 (CONFORMALITY, OR THE PRESERVATION OF ANGLES).   *Let* $S^n \subset \mathbb{R}^{n+1}$, $P$, $S$, $N$, *and* $\pi$ *(extended) be as in the definition. Then* $\pi$ *preserves angles between curves in* $S^n \setminus \{N\}$. *Furthermore, if* $x \in S^n \setminus \{N, S\}$ *and if* $T = xy$ *is a line segment tangent to* $S^n$ *at* $x$, *then the angles* $\pi(x)\,x\,y$ *and* $x\,\pi(x)\,\pi(y)$ *are either equal or complementary whenever* $\pi(y)$ *is defined.*

PROOF. We first give the analytic proof in arbitrary dimensions that $\pi$ preserves angles between curves in $S^n \setminus \{N\}$.

We may clearly normalize everything so that $S^n$ is in fact the unit sphere in $\mathbb{R}^{n+1}$, $S$ is the point with coordinates $(0, \ldots, 0, -1)$, $N$ is the point with coordinates $(0, \ldots, 0, 1)$, $P$ is the plane $x_{n+1} = -1$, and $\pi : S^n \to P$ is given by the formula $\pi(x) = (y_1, \ldots, y_n, -1)$, where

$$y_i = \frac{-2}{x_{n+1} - 1} x_i.$$

We take the Euclidean metric $ds^2 = dy_1^2 + \cdots + dy_n^2$ on $P$ and pull it back to a metric $\pi^*(ds^2)$ on $S^n$. The pullback of $dy_i$ is the form

$$\frac{-2}{x_{n+1} - 1}\left(dx_i - \frac{x_i}{x_{n+1} - 1} dx_{n+1}\right).$$

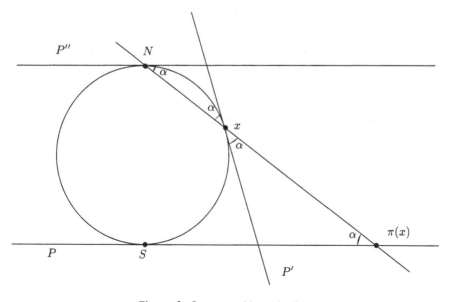

**Figure 6.** Stereographic projection.

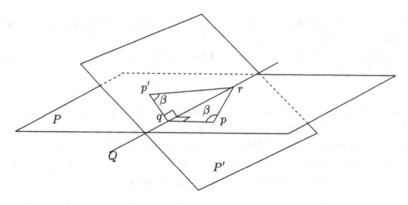

**Figure 7.** The angles $qpr$ and $qp'r$.

Because $x \in S^n$, we have the two equations

$$x_1^2 + \cdots + x_n^2 + x_{n+1}^2 = 1$$

and

$$x_1 \, dx_1 + \cdots + x_n \, dx_n + x_{n+1} \, dx_{n+1} = 0.$$

From these equations it is easy to deduce that

$$\pi^*(ds^2) = \frac{4}{(x_{n+1} - 1)^2}(dx_1^2 + \cdots + dx_n^2 + dx_{n+1}^2);$$

the calculation is essentially identical with one that we have performed above. We conclude that at each point the pullback of the Euclidean metric on $P$ is a positive multiple of the Euclidean metric on $S^n$. Since multiplying distances in a tangent space by a positive constant does not change angles, the map $\pi : S^n \setminus \{N\} \to P$ preserves angles.

For the second assertion of the theorem we give a geometric proof that, in the special case of dimension $n + 1 = 3$, also gives an alternative geometric proof of the fact that we have just proved analytically. This proof is taken from [Hilbert and Cohn-Vossen 1932].

In preparation we consider two planes $P$ and $P'$ of dimension $n$ in Euclidean $(n + 1)$-space $\mathbb{R}^{n+1}$ that intersect in a plane $Q$ of dimension $n - 1$. We then pick points $p \in P$, $q \in Q$, and $p' \in P'$ such that the line segments $pq$ and $p'q$ are of equal length and are at right angles to $Q$.

As can be seen in Figure 7, if $r \in Q$, the angles $qpr$ and $qp'r$ are equal. Similarly, the angles $p'pr$ and $pp'r$ are equal.

To prove the second assertion of the theorem, first note that the case in which the line $M$ containing $x$ and $y$ misses $P$ follows by continuity from the case in which $M$ meets $P$. So suppose that $M$ meets $P$. Note that $\pi$ maps the points of $M$ for which $\pi$ is defined to the line containing $\pi(x)$ and $\pi(y)$. This implies that we may assume that $y \in P$. See Figure 6. Now for the plane $P$ of the obvious assertion we take the plane $P$ tangent to the sphere $S^n$ at the south pole $S$. For the plane $P'$ of the obvious assertion we take the plane tangent to $S^n$ at $x$. For the points $p' \in P'$ and $p \in P$ we take, respectively, the points $p = \pi(x) \in P$ and $p' = x \in P'$. For the plane $Q$ we take the intersection of $P$ and $P'$. For the point $r$ we take $y$. Now the assertion that the angles $p'pr$ and $pp'r$ are equal proves the second assertion of the theorem.

In dimension three, the obvious fact that the angles $qpr$ and $qp'r$ are equal shows that $\pi$ preserves the angle between any given curve and certain reference tangent directions, namely $pq$ and $p'q$. Since the tangent space is, in this dimension only, two-dimensional, preserving angle with reference tangent directions is enough to ensure preservation of angle in general.                                               □

THEOREM 8.2 (PRESERVATION OF SPHERES). *Assume the setting of the previous theorem. If $C$ is a sphere ($C$ for circle) in $S^n$ that passes through the north pole $N$ of $S^n$ and has dimension $c$, then the image $\pi(C) \subset P$ is a plane in $P$ of dimension $c$. If on the other hand $C$ misses $N$, then the image $\pi(C)$ is a sphere in $P$ of dimension $c$.*

PROOF. If $N \in C$, then the proof is easy; indeed $C$ is contained in a unique plane $P'$ of dimension $c+1$, and the image $\pi(C)$ is the intersection of $P'$ and $P$, a $c$-dimensional plane.

If, on the other hand, $C$ misses $N$, we argue as follows. We assume all normalized as in the analytic portion of the proof of the previous theorem so that $S^n$ is the unit sphere. We can deal with the case where $C$ is a union of great circles by continuity if we manage to prove the theorem in all other cases. Consequently, we may assume that the vector subspace of $\mathbb{R}^{n+1}$ spanned by the vectors in $C$ has dimension $c + 2$. We lose no generality in assuming that it is all of $\mathbb{R}^{n+1}$ (that is, $c = n - 1$).

The tangent spaces to $S^n$ at the points of $C$ define a conical envelope with cone point $y$; one easy way to find $y$ is to consider the two-dimensional plane $R$ containing $N$ and two antipodal points $r$ and $r'$ of $C$, and to consider the two tangent lines $t(r)$ to $C \cap R$ at $r$ and $t(r')$ to $C \cap R$ at $r'$; then $y$ is the point at which $t(r)$ and $t(r')$ meet. See Figure 8. By continuity we may assume that $\pi(y)$ is defined.

We assert that $\pi(y)$ is equidistant from the points of $\pi(C)$, from which the reader may deduce that $\pi(C)$ is a sphere centered at $\bar{\pi}(y)$. By continuity it suffices to prove that $\pi(y)$ is equidistant from the points of $\pi(C) \setminus S$. Here is the argument that proves the assertion. Let $x \in C \setminus S$, and consider the two-dimensional plane containing $N$, $x$, and $y$. In this plane there is a point $x'$ on the line through $x$ and $N$ such that the line segment $yx'$ is parallel to the segment $\pi(y)\pi(x)$; that is, the angles $N\,\pi(x)\,\pi(y)$ and $N\,x'\,y$ are equal. By the final assertion of Theorem 8.1, the angles $\pi(y)\,\pi(x)\,x$ and $y\,x\,\pi(x)$ are either equal or complementary. Thus the triangle $xyx'$ is isosceles so that sides $xy$ and $x'y$ are equal. Thus considering proportions in the similar triangles $N\,x'\,y$ and

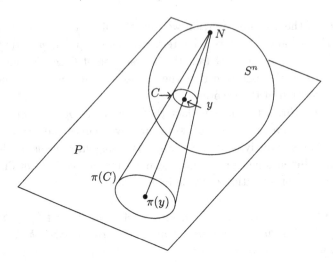

**Figure 8.** Stereographic projection maps spheres to spheres.

$N\,\pi(x)\,\pi(y)$, we have the equalities

$$d(\pi(x), \pi(y)) = \frac{d(N, \pi(y))}{d(N, y)}\, d(x', y) = \frac{d(N, \pi(y))}{d(N, y)}\, d(x, y).$$

Of course, the fraction is a constant since $N$, $y$, and $\pi(y)$ do not depend on $x$; and the distance $d(x, y)$ is also a constant since $x \in C$, $C$ is a sphere, and $y$ is the center of the tangent cone of $C$. We conclude that the distance $d(\pi(x), \pi(y))$ is constant.                                               □

DEFINITION. Let $S^n$ denote a sphere of dimension $n$ in $\mathbb{R}^{n+1}$ with north pole $N$ and south pole $S$ as above. Let $P$ denote a plane through the center of $S^n$ and orthogonal to the line through $N$ and $S$. If $x$ is any point of $S^n \setminus \{N\}$, then there is a unique point $\pi'(x)$ of $P$ on the line that contains $N$ and $x$. This defines a map $\pi' : S^n \setminus \{N\} \to P$, *stereographic projection* from $S^n \setminus \{N\}$ to $P$.

THEOREM 8.3. *The map $\pi'$ preserves angles between curves in $S^n \setminus \{N\}$, and $\pi'$ maps spheres to planes or spheres.*

PROOF. We normalize so that $S^n$ is the unit sphere in $\mathbb{R}^{n+1}$, $N = (0, \ldots, 0, 1)$, and $S = (0, \ldots, 0, -1)$. From the proof of Theorem 8.1 we have for every $x \in S^n \setminus \{N\}$ that $\pi(x) = (y_1, \ldots, y_n, -1)$, where

$$y_i = \frac{-2}{x_{n+1} - 1}\, x_i.$$

In the same way $\pi'(x) = (y_1', \ldots, y_n', -1)$, where

$$y_i' = \frac{-1}{x_{n+1} - 1}\, x_i = \frac{y_i}{2}.$$

Thus $\pi'$ is the composition of $\pi$ with a translation and a dilation. Since $\pi$ preserves angles and maps spheres to planes or spheres, so does $\pi'$.          □

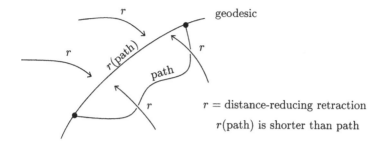

$r$ = distance-reducing retraction
$r$(path) is shorter than path

**Figure 9.** The retraction principle.

## 9. Geodesics

Having established formulas for the hyperbolic metric in our five analytic models and having developed the fundamental properties of stereographic projection, it is possible to find the straight lines or *geodesics* in our five models with a minimal amount of effort. Though geodesics can be found by solving differential equations, we shall not do so. Rather, we establish the existence of one geodesic in the half-space model by means of what we call the retraction principle. Then we deduce the nature of all other geodesics by means of simple symmetry properties of the hyperbolic metrics. Here are the details. We learned this argument from Bill Thurston.

THEOREM 9.1 (THE RETRACTION PRINCIPLE). *Suppose that $X$ is a Riemannian manifold, that $C : (a, b) \to X$ is an embedding of an interval $(a, b)$ in $X$, and that there is a retraction $r : X \to$ image$(C)$ that is distance-reducing in the sense that, if one restricts the metric of $X$ to image$(C)$ and pulls this metric back via $r$ to obtain a new metric on all of $X$, then at each point the pullback metric is less than or equal to the original metric on $X$. Then the image of $C$ contains a shortest path (geodesic) between each pair of its points.*

The proof is left as an exercise. (Take an arbitrary path between two points of the image and show that the retraction of that path is at least as short as the original path. See Figure 9.)

THEOREM 9.2 (EXISTENCE OF A FUNDAMENTAL GEODESIC IN HYPERBOLIC SPACE). *In the half-space model of hyperbolic space, all vertical lines are geodesic. Such a line is the unique shortest path between any pair of points on it.*

PROOF. Let $C : (0, \infty) \to H$, where $C(t) = (1, x_2, \ldots, x_n, t) \in H$ and where the numbers $x_2, \ldots, x_n$ are fixed constants; that is, $C$ is an arbitrary vertical line in $H$.

Define a retraction $r : H \to$ image$(C)$ by the formula

$$r(1, x'_1, \ldots, x'_n, t) = (1, x_1, \ldots, x_n, t).$$

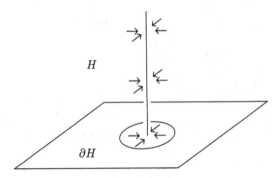

**Figure 10.** A fundamental hyperbolic geodesic and a distance-reducing retraction

See Figure 10. The original hyperbolic metric was $ds^2 = (dx_2^2 + \cdots + dx_{n+1}^2)/x_{n+1}^2$. The pullback metric is $dx_{n+1}^2/x_{n+1}^2$. Thus, by the retraction principle, the image of $C$ contains a shortest path between each pair of its points.

It remains only to show that there is only one shortest path between any pair of points on the image of $C$. If one were to start with an arbitrary path between two points of the image of $C$ that does not in fact stay in the image of $C$, then at some point the path is not vertical; hence the pullback metric is actually smaller than the original metric at that point since the original metric involves some $dx_i^2$ with $i \neq n + 1$. Thus the retraction is actually strictly shorter than the original path. It is clear that there is only one shortest path between two points of the image that stays in the image. □

THEOREM 9.3 (CLASSIFICATION OF GEODESICS IN $H$). *The geodesics in the half-space model $H$ of hyperbolic space are precisely the vertical lines in $H$ and the Euclidean metric semicircles whose endpoints lie in and intersect the boundary $\{(1, x_2, \ldots, x_n, 0)\}$ of hyperbolic space $H$ orthogonally.*

PROOF. See Figure 11 for the two types of geodesics. We need to make the following observations:

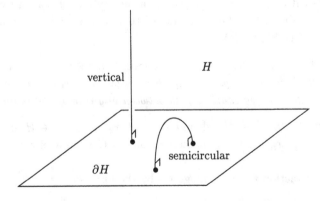

**Figure 11.** The two types of geodesics in $H$.

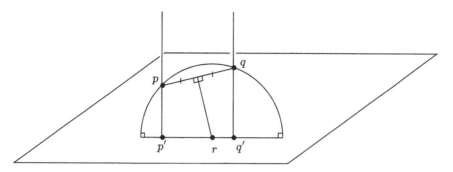

**Figure 12.** Finding the hyperbolic geodesic between points of $H$ not on a vertical line.

(1) Euclidean isometries of $H$ that take the boundary $\{(1, x_2, \ldots, x_n, 0)\}$ of $H$ to itself are hyperbolic isometries of $H$. Similarly, the transformations of $H$ that take $(1, x_1, \ldots, x_n, t)$ to $(1, rx_1, \ldots, rx_n, rt)$ with $r > 0$ are hyperbolic isometries. (Proof by direct, easy calculation.)

(2) Euclidean isometries of $J$ are hyperbolic isometries of $J$. (Proof by direct, easy calculation.)

(3) If $p$ and $q$ are arbitrary points of $H$, and if $p$ and $q$ do not lie on a vertical line, then there is a unique boundary-orthogonal semicircle that contains $p$ and $q$. Indeed, to find the center of the semicircle, take the Euclidean segment joining $p$ and $q$ and extend its Euclidean perpendicular bisector in the vertical plane containing $p$ and $q$ until it touches the boundary of $H$. See Figure 12.

(4) If $C$ and $C'$ are any two boundary-orthogonal semicircles in $H$, then there is a hyperbolic isometry taking $C$ onto $C'$. (The proof is an easy application of (1) above.)

We now complete the proof of the theorem as follows. By the previous theorem and (1), all vertical lines in $H$ are geodesic and hyperbolically equivalent, and each contains the unique shortest path between each pair of its points. Now map the vertical line in $H$ with infinite endpoint $(1, 0, \ldots, 0)$ into $J$ via the connecting stereographic projection. Then the image is a great semicircle. Rotate $J$, a hyperbolic isometry by (2), so that the center of the stereographic projection is not an infinite endpoint of the image. Return the rotated semicircle to $H$ via stereographic projection. See Figure 13. By the theorems on stereographic projection, the image is a boundary-orthogonal semicircle in $H$. Since it is the image under a composition of isometries of a geodesic, this boundary-orthogonal semicircle is a geodesic. But all boundary-orthogonal semicircles in $H$ are hyperbolically equivalent by (4) above. Hence each is a geodesic. Since there is a unique geodesic joining any two points of a vertical line, we find that there is a unique geodesic joining any two points of $H$ (see (3)). This completes the proof of the theorem. $\qquad\square$

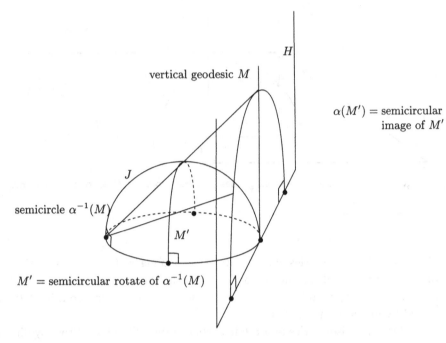

**Figure 13.** Geodesics in $H$.

By Theorems 8.1 and 8.2, the boundary-orthogonal semicircles in $J$ correspond precisely to the boundary-orthogonal semicircles and vertical lines in $H$. Hence the geodesics in $J$ are the boundary-orthogonal semicircles in $J$.

By Theorem 8.3, the boundary-orthogonal semicircles in $J$ correspond to the diameters and boundary-orthogonal circular segments in $I$. Hence the diameters and boundary-orthogonal circular segments in $I$ are the geodesics in $I$. See Figure 14.

The boundary-orthogonal semicircles in $J$ clearly correspond under vertical projection to straight line segments in $K$. Hence the latter are the geodesics in $K$. See Figure 15.

The straight line segments in $K$ clearly correspond under central projection from the origin to the intersections with $L$ of two-dimensional vector subspaces of $\mathbb{R}^{n+1}$ with $L$; hence the latter are the geodesics of $L$. See Figure 16.

## 10. Isometries and Distances in the Hyperboloid Model

We begin our study of the isometries of hyperbolic space with the hyperboloid model $L$ where all isometries, as we shall see, are restrictions of linear maps of $\mathbb{R}^{n+1}$.

DEFINITION. A *linear isometry* $f : L \to L$ of $L$ is the restriction to $L$ of a linear map $F : \mathbb{R}^{n+1} \to \mathbb{R}^{n+1}$ that preserves the hyperbolic inner product $*$ (that is,

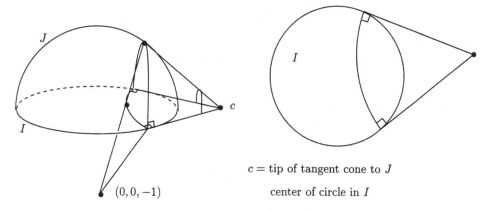

$c =$ tip of tangent cone to $J$

center of circle in $I$

**Figure 14.** Geodesics in $I$ and $J$ and their stereographic relationship.

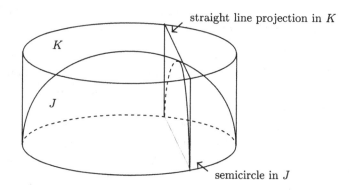

**Figure 15.** Geodesics in $J$ and $K$.

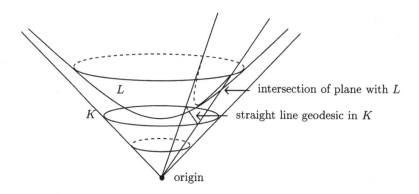

**Figure 16.** Geodesics in $K$ and $L$.

for each pair $v$ and $w$ of vectors from $\mathbb{R}^{n+1}$, $Fv * Fw = v * w$) and that takes the upper sheet of the hyperboloid $L$ into itself.

DEFINITION. A *Riemannian isometry* $f : L \to L$ of $L$ is a diffeomorphism of $L$ that preserves the Riemannian metric (that is, $f^*(ds^2) = ds^2$).

DEFINITION. A *topological isometry* $f : L \to L$ of $L$ is a homeomorphism of $L$ that preserves the Riemannian distance between each pair of points of $L$ (that is, if $d$ is the Riemannian distance function and if $x$ and $y$ are points of $L$, then $d(f(x), f(y)) = d(x, y)$).

THEOREM 10.1. *A square matrix $M$ with columns $m_1, \ldots, m_n, m_{n+1}$ induces a linear isometry of $L$ if and only if the following two conditions are satisfied.*

1. *For each pair of indices $i$ and $j$, we have $m_i * m_j = e_i * e_j$, where $e_1, \ldots, e_n$, $e_{n+1}$ is the standard basis for $\mathbb{R}^{n+1}$.*
2. *The last entry of the last column $m_{n+1}$ is positive.*

*Condition 1 is satisfied if and only if $M$ is invertible with $M^{-1} = JM^tJ$, where $J$ is the diagonal matrix with diagonal entries $J_{11} = \cdots = J_{nn} = -J_{n+1,n+1} = 1$.*

PROOF. Let $J$ denote the diagonal matrix with diagonal entries $J_{11} = \cdots = J_{nn} = -J_{n+1,n+1} = 1$. Then for each $x, y \in \mathbb{R}^{n+1}$, $x*y = x^t Jy$. Thus $Mx*My = x^t M^t JMy$. Consequently, $M$ preserves $*$ if and only if $M^t JM = J$; but the $ij$ entry of $M^t JM$ is $m_i * m_j$ while that of $J$ is $e_i * e_j$. Thus $M$ preserves $*$ if and only if condition 1 of the theorem is satisfied. Since $J$ is invertible, condition 1 implies that $M$ is also invertible and that it takes the hyperboloid of two sheets, of which $L$ is the upper sheet, homeomorphically onto itself. Condition 2 is then just the statement that the image of $e_{n+1}$ lies in $L$, that is, that $M$ takes the upper sheet $L$ of the hyperboloid onto itself.

Finally, the equality $M^{-1} = JM^tJ$ is clearly equivalent to the equality $M^t JM = J$ since $J^{-1} = J$. $\qquad \square$

THEOREM 10.2. *A map $f : L \to L$ that satisfies any of the three definitions of isometry—linear, Riemannian, or topological—satisfies the other two as well.*

PROOF. We first prove the two easy implications, linear $\Rightarrow$ Riemannian $\Rightarrow$ topological, then connect the hyperbolic inner product $x * y$ with Riemannian distance $d(x, y)$ in preparation for the more difficult implication, topological $\Rightarrow$ linear.

*Linear isometry $\Rightarrow$ Riemannian isometry:* Let $F : \mathbb{R}^{n+1} \to \mathbb{R}^{n+1}$ be a linear map that preserves the hyperbolic inner product $*$ and takes the upper sheet $L$ of the hyperboloid of two sheets into itself and thereby induces a linear isometry $f : L \to L$. The Riemannian metric $ds^2$ is at each point $x$ of $L$ simply a function of two variables that takes as input two tangent vectors $v$ and $w$ at $x$ and delivers as output the hyperbolic inner product $v * w$. We calculate the pullback metric

$f^*(ds^2)$ in the following manner:

$$f^*(ds^2)(v,w) = ds^2(Df(v), Df(w)) = ds^2(DF(v), DF(w))$$
$$= ds^2(F(v), F(w)) = F(v) * F(w) = v * w = ds^2(v,w).$$

We conclude that $f^*(ds^2) = ds^2$, so that $f$ is a Riemannian isometry.

*Riemannian isometry $\Rightarrow$ topological isometry:* Riemannian distance is calculated by integrating the Riemannian metric. Since a Riemannian isometry preserves the integrand, it preserves the integral as well.

LEMMA. *If $a, b \in L$, then $a * b = -\cosh(d(a,b))$.*

PROOF. Let $t$ denote the Riemannian distance $d(a,b)$ between $a$ and $b$. One obtains this distance by integrating the Riemannian metric along the unique geodesic path joining $a$ and $b$, or, since this integral is invariant under linear isometry, one can translate $a$ and $b$ to a standard position in $L$ as follows and then perform the integration. Let $m_1$ be the unit tangent vector at $a$ in the direction of the geodesic from $a$ to $b$. Let $m_{n+1} = a$. By the Gram–Schmidt orthonormalization process from elementary linear algebra we may extend the orthonormal set $\{m_1, m_{n+1}\}$ to an orthonormal basis $m_1, \ldots, m_n, m_{n+1}$ for $\mathbb{R}^{n+1}$; that is, $m_i * m_j = e_i * e_j$. By Theorem 10.1, the matrix $M$ with columns $m_1, \ldots, m_n, m_{n+1}$ gives a linear isometry of $L$ as does its inverse $M^{-1}$. The inverse takes $a$ to $e_{n+1}$ and takes the two-dimensional subspace spanned by $a$ and $b$ to the space $P$ spanned by $e_1$ and $e_{n+1}$. The intersection of $P$ with $L$ is one branch of a standard hyperbola that passes through $M^{-1}(a)$ and $M^{-1}(b)$ and is the unique hyperbolic geodesic through those two points. Since $M^{-1}(a) = (0, \ldots, 0, 1)$ and since $t = d(a,b) = d(M^{-1}(a), M^{-1}(b))$, we may assume that $M^{-1}(b) = (\sinh(t), \ldots, 0, \cosh(t))$. (See Section 4.) Thus we may calculate:

$$a * b = M^{-1}(a) * M^{-1}(b) = (0, \ldots, 0, 1) * (\sinh(t), \ldots, 0, \cosh(t))$$
$$= -\cosh(t) = -\cosh(d(a,b)). \qquad \square$$

*Topological isometry $\Rightarrow$ linear isometry:* Let $f : L \to L$ denote a topological isometry. Let $v_1, \ldots, v_n, v_{n+1}$ denote a basis for $\mathbb{R}^{n+1}$ such that each $v_i$ lies in $L$. Let $F$ denote the linear map that takes $v_i$ to $f(v_i)$ for each $i$. We claim that $F$ preserves $*$; to see this, write $e_i = \sum_j a_{ij} v_j$ and compute:

$$F(e_i) * F(e_j) = \sum_{k,l} a_{ik} a_{jl} f(v_k) * f(v_l)$$
$$= \sum_{k,l} a_{ik} a_{jl} (-\cosh(d(f(v_k), f(v_l))))$$
$$= \sum_{k,l} a_{ik} a_{jl} (-\cosh(d(v_k, v_l))) = e_i * e_j.$$

Moreover, $F$ agrees with $f$ on $L$. To prove this, it suffices to replace $f$ by $F^{-1} \circ f$, so that we can assume $f(v_i) = v_i$; then we must prove that $f = \mathrm{id}$, which we can do by showing $f(x) * e_i = x * e_i$ for each $x \in L$ and for each index $i$. Here is the calculation:

$$f(x) * e_i = f(x) * \sum_j a_{ij} v_j = \sum_j a_{ij} (f(x) * f(v_j))$$

$$= \sum_j a_{ij} (- \cosh(d(f(x), f(v_j))))$$

$$= \sum_j a_{ij} (- \cosh(d(x, v_j))) = x * e_i. \qquad \square$$

## 11. The Space at Infinity

It is apparent from all of our analytic models with the possible exception of the hyperboloid model $L$ that there is a natural space at infinity. In the half-space model $H$ it is the bounding plane of dimension $n - 1$ that we compactify by adding one additional point; we visualize the additional point as residing at the top of the collection of vertical geodesics in $H$. In the disk model $I$, in the hemisphere model $J$, and in the Klein model $K$ it is the bounding $(n-1)$-sphere. If we reinterpret the hyperboloid model as lying in projective space (each point of $L$ is represented by the unique one-dimensional vector subspace of $\mathbb{R}^{n+1}$ that contains that point), then the space at infinity becomes apparent in that model as well: it consists of those lines that lie in the light cone $\{x \in \mathbb{R}^{n+1} : x * x = 0\}$. Furthermore, it is apparent that not only the models but also the unions of those models with their spaces at infinity correspond homeomorphically under our transformations connecting the models. That is, the space at infinity is a sphere of dimension $n - 1$ and the union of the model with the space at infinity is a ball of dimension $n$.

Having analyzed the isometries of the hyperboloid model, we see that each isometry of $L$ actually extends naturally not only to the space at infinity but to the entirety of projective $n$-space. That is, each linear mapping of $\mathbb{R}^{n+1}$ defines a continuous mapping of projective $n$-space $P^n$.

## 12. The Geometric Classification of Isometries

We recall from the previous sections that every isometry $f$ of $L$ extends to a linear homeomorphism $F$ of $\mathbb{R}^{n+1}$, hence upon passage to projective space $P^n$ induces a homeomorphism $f \cup f_\infty : L \cup \partial L \to L \cup \partial L$ of the ball that is the union of hyperbolic space $L$ and its space $\partial L$ at infinity. Every continuous map from a ball to itself has a fixed point by the Brouwer fixed point theorem. There is a very useful and beautiful geometric classification of the isometries of hyperbolic space that refers to the fixed points of this extended map. Our analysis of these maps requires that we be able to normalize them to some extent by moving given fixed points into a standard position. To that end we note that we have already shown how to move any point in $L$ and nonzero tangent vector at that point so that the point is at $e_{n+1}$ and the tangent points in the direction of $e_1$. As a consequence we can move any pair of points in $L \cup \partial L$ so that they lie in any given geodesic; and by conjugation we find that we may assume that any pair of fixed points of an isometry lies in a given geodesic. Indeed, let $f$ be an isometry with fixed point $x$, let $g$ be an isometry that takes $x$ into a geodesic line $L$, and note that $g(x)$ is a fixed point of $gfg^{-1}$. Here are the three possible cases.

*The elliptic case* occurs when the extended map has a fixed point in $L$ itself: conjugating by a linear isometry of $L$, we may assume that the isometry $f$ : $L \to L$ fixes the point $e_{n+1} = (0, \ldots, 0, 1)$. Let $F : \mathbb{R}^{n+1} \to \mathbb{R}^{n+1}$ be the linear extension of $f$. The representing matrix $M$ has as last column $m_{n+1}$ the vector $e_{n+1}$. The remaining columns must be $*$-orthogonal to $m_{n+1}$, hence Euclidean or $\cdot$-orthogonal to $e_{n+1}$. On the orthogonal complement of $e_{n+1}$, the hyperbolic and the Euclidean inner products coincide. Hence the remaining columns form not only a hyperbolic orthonormal basis but also a Euclidean orthonormal basis. We conclude that the matrix $M$ defining $F$ is actually Euclidean orthogonal. We call such a transformation of hyperbolic space *elliptic*.

*The hyperbolic case* occurs when the extended map has no fixed point in $L$ itself but has two fixed points at infinity: we examine this transformation in the half-space model $H$ for hyperbolic space. We ignore the initial constant coordinate 1 in $H$ and identify $H$ with the half-space $\{x = (x_1, \ldots, x_n) \in \mathbb{R}^n : x_n > 0\}$. Conjugating by an isometry, we may assume that the fixed points of the map $f$ of $H \cup \partial H$ are the infinite endpoints of the hyperbolic geodesic $(0, \ldots, t)$, where $t > 0$. Let $(0, \ldots, k)$ denote the image under $f$ of $(0, \ldots, 1)$. Then $(1/k) \cdot f$ is an isometry that fixes every point of the hyperbolic geodesic $(0, \ldots, t)$. By the previous paragraph, the transformation $(1/k) \cdot f$ is an orthogonal transformation $O$. It follows easily that $f(x) = k\, O(x)$, the composite of a Euclidean orthogonal transformation $O$, which preserves the boundary plane at infinity and which is simultaneously a hyperbolic isometry, with the hyperbolic translation $x \mapsto kx$ along the geodesic $(0, \ldots, t)$. Such a transformation is called *hyperbolic* or *loxodromic*. The invariant geodesic $(0, \ldots, t)$ is called the *axis* of the hyperbolic transformation. See Figure 17. Often one preserves the name *hyperbolic* for the case where the orthogonal transformation is trivial and the name *loxodromic* for the case where the orthogonal transformation is nontrivial.

*The parabolic case* occurs when the extended map has only one fixed point and that fixed point is at infinity: we examine this transformation in the half-space model $H$ for hyperbolic space. We may assume that the fixed point of the map $f$ of $H \cup \partial H$ is the upper infinite endpoint of the hyperbolic geodesic $(0, \ldots, t)$, where $t > 0$. The transformation $g : x \mapsto f(x) - f((0, \ldots, 0))$ fixes both ends of the same geodesic. Hence $g$ may be written as a composite $x \mapsto k\, O(x)$ where $k > 0$ and $O$ have the significance described in the previous paragraph. Thus $f(x) = k\, O(x) + v$, where $k > 0$, $O$ is Euclidean orthogonal preserving the boundary plane of $H$, and $v = f((0, \ldots, 0))$ is a constant vector. We claim that

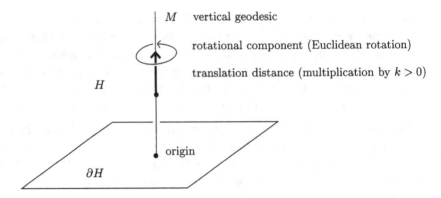

**Figure 17.** Hyperbolic or loxodromic isometry.

$k = 1$ so that $f$ is a Euclidean isometry preserving the boundary plane of $H$; such a map, without fixed points in the boundary plane, is called *parabolic*. If $k \neq 1$, we claim that $f$ has another fixed point. We find such a fixed point in the following way. The fixed point will be a solution of the equation $(I - kO)x = v$. The eigenvalues of $I - kO$ have the form $1 - k\lambda$, where $\lambda$ is an eigenvalue of $O$. Since $O$ is orthogonal, its eigenvalues have absolute value 1. Hence if $k \neq 1$, then $I - kO$ is invertible, and the equation $(I - kO)x = v$ does indeed have a solution.

## 13. Curious Facts about Hyperbolic Space

FACT 1. *In the three conformal models for hyperbolic space, hyperbolic spheres are also Euclidean spheres; however, Euclidean and hyperbolic sphere centers need not coincide.*

PROOF. We work in the hemisphere model $J$ for hyperbolic space and consider the point $p = (0, \dots, 0, 1) \in J$. The Riemannian metric $ds_J^2$ is clearly rotationally symmetric around $p$ so that a hyperbolic sphere centered at $p$ is a Euclidean sphere.

We project such a sphere from $J$ into the half-space model $H$ for hyperbolic space via stereographic projection. See Figure 18. Since stereographic projection takes spheres that miss the projection point to spheres in $H$, we see that there is one point of $H$, namely the image of $p$, about which hyperbolic spheres are Euclidean spheres. But this point can be taken to any other point of $H$ by a composition of Euclidean translations and Euclidean similarities that are hyperbolic isometries as well. Since these Euclidean transformations preserve both the class of hyperbolic spheres and the class of Euclidean spheres, we see that the hyperbolic spheres centered at each point of $H$ are also Euclidean spheres.

We project this entire class of spheres back into $J$ and from thence into $I$ by stereographic projections that preserve this class of Euclidean (and hyperbolic)

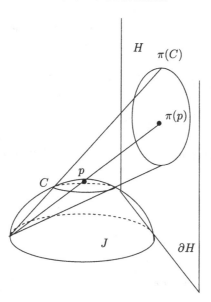

**Figure 18.** The projection of a sphere from $J$ to $H$. $C$ is both a Euclidean and a hyperbolic sphere; $\pi(p)$ is the hyperbolic center of the circle $\pi(C)$.

spheres. We conclude that all hyperbolic spheres in these three models are also Euclidean spheres, and conversely.

Finally, we give a geometric construction for the hyperbolic center of a Euclidean sphere $S$ in the half-space model $H$. See Figure 19. Draw the vertical geodesic line $M$ through the center of $S$ until it meets the plane at infinity at some point $p$. Draw a tangent line to $S$ from $p$ meeting $S$ at a tangency point $q$. Draw the circle $C$ through $q$ that is centered at $p$ and lies in the same plane as $M$. The circle $C$ then meets the line $M$ at the hyperbolic center of $S$ (proof, an exercise for the reader). Note that this center is not the Euclidean center of $S$. $\qquad\square$

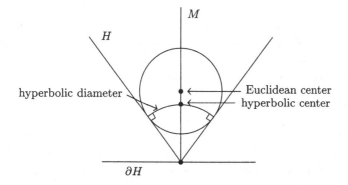

**Figure 19.** Constructing the hyperbolic center of a circle

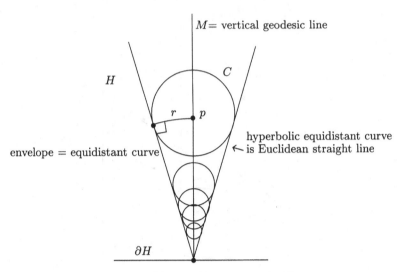

**Figure 20.** Equidistant curves in $H$.

FACT 2. *In the hyperbolic plane, the two curves at distance $r$ on either side of a straight line are not straight.*

PROOF. We can use the preceding result to analyze the curves equidistant from a hyperbolic geodesic in the hyperbolic plane. We work in the half-space model $H \subset \mathbb{R}^2$ of the two-dimensional hyperbolic plane and take as geodesic line the vertical line $M$ that passes through the origin of $\mathbb{R}^2$. Put a hyperbolic circle $C$ of hyperbolic radius $r$ about a point $p$ of $M$. Then we obtain the set of all such circles centered at points of $M$ by multiplying $C$ by all possible positive scalars. The union of these spheres $t \cdot C$ is a cone, or angle, $D$ of which the origin is the vertex and whose central axis is $M$. The envelope or boundary of this cone or angle is a pair of Euclidean straight lines, the very equidistant lines in which we are interested. See Figure 20. Since these straight lines are not vertical, they are not hyperbolic straight lines. □

FACT 3. *Triangles in hyperbolic space have angle sum less than $\pi$; in fact, the area of a triangle with angles $\alpha$, $\beta$, and $\gamma$ is $\pi - \alpha - \beta - \gamma$ (the Gauss–Bonnet theorem). Given three angles $\alpha$, $\beta$, and $\gamma$ whose sum is less than $\pi$, there is one and only one triangle up to congruence having those angles. Consequently, there are no nontrivial similarities of hyperbolic space.*

PROOF. Any triangle in hyperbolic space lies in a two-dimensional hyperbolic plane. Hence we may work in the half-space model $H$ for the hyperbolic plane. Assume that we are given a triangle $\Delta = pqr$ with angles $\alpha$, $\beta$, and $\gamma$. We may arrange via an isometry of hyperbolic space that the side $pq$ lies in the unit circle. Then by a hyperbolic isometry of hyperbolic space that has the unit circle as its invariant axis and translates along the unit circle we may arrange that the side $pr$ points vertically upward. The resulting picture is in Figure 21.

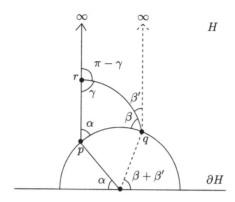

**Figure 21.** The Gauss–Bonnet theorem.

We note that the triangle $\Delta = pqr$ is the difference of two ideal triangles $pq\infty$ and $rq\infty$. ("Ideal" means that at least one vertex is at infinity.) We first prove the Gauss–Bonnet theorem for such an ideal triangle, then deduce the desired formula by taking a difference.

The element of area is $dA = dx\,dy/y^2$. It is easy to verify that the area of $pq\infty$ is therefore

$$\int_{pq\infty} dA = \int_{x=\cos(\beta+\beta')}^{x=\cos(\pi-\alpha)} dx \int_{y=\sqrt{1-x^2}}^{y=\infty} \frac{dy}{y^2}.$$

Straightforward evaluation leads to the value $\pi - \alpha - \beta - \beta'$ for the integral. Similar evaluation gives the value $\pi - (\pi - \gamma) - \beta'$ for the area of $rq\infty$. The difference of these two values is $\pi - \alpha - \beta - \gamma$ as claimed. This proves the Gauss–Bonnet theorem.

We now construct a triangle with given angles. Suppose therefore that three angles $\alpha$, $\beta$, and $\gamma$ are given whose sum is less than $\pi$, a necessary restriction in view of the Gauss–Bonnet theorem. Pick a model of the hyperbolic plane, say the disk model $I$. Pick a pair $Q$ and $R$ of geodesic rays (radii) from the origin $p$ meeting at the Euclidean ($=$ hyperbolic) angle of $\alpha$. See Figure 22. Note that any pair of geodesic rays meeting at angle $\alpha$ is congruent to this pair. Pick points $q$ and $r$ on these rays and consider the triangle $pqr$. Let $\beta'$ denote the angle at $q$ and let $\gamma'$ denote the angle at $r$. Let $A'$ denote the area of the triangle $pqr$. We will complete the construction by showing that there is a unique choice for $q$ on $Q$ and for $r$ on $R$ such that $\beta' = \beta$ and $\gamma' = \gamma$. The argument will be variational.

We first consider the effect of fixing a value of $q$ and letting $r$ vary from $\infty$ to $p$ along $R$. At $\infty$, the angle $\gamma'$ is 0. At (near) $p$ the angle $\gamma'$ is (almost) $\pi - \alpha$. As $r$ moves inward toward $p$ along $R$, both $\beta'$ and $A'$ clearly decrease monotonically. Hence, by the Gauss–Bonnet theorem, $\gamma' = \pi - \alpha - \beta' - A'$ increases monotonically. In particular, there is a unique point $r(q)$ at which

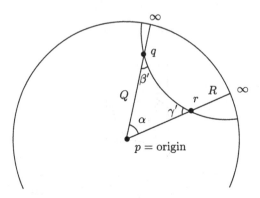

**Figure 22.** Constructing a triangle with angles $\alpha$, $\beta$, and $\gamma$, with $\alpha + \beta + \gamma < \pi$

$\gamma' = \gamma$. Now fix $q$, fix $r$ at $r(q)$, and move inward along $Q$ from $q$ to a point $q'$. Note that the angle of $pq'r$ at $r$ is smaller than the angle $\gamma$, which is the angle of $pqr$ at $r$. We conclude that $r(q')$ must be closer to $p$ than is $r(q)$. That is, as $q$ moves inward toward $p$, so also does $r(q)$. We conclude that the areas of the triangles $pqr(q)$ decrease monotonically as $q$ moves inward along $Q$ toward $p$.

We are ready for the final variational argument. We work with the triangles $pqr(q)$. We start with $q$ at $\infty$ and note that the area $A'$ is equal to $\pi - \alpha - 0 - \gamma > \pi - \alpha - \beta - \gamma$. We move $q$ inward along $Q$, and consequently move $r(q)$ inward along $R$, until $q$ reaches $p$ and $A' = 0 < \pi - \alpha - \beta - \gamma$. As noted in the previous paragraph, the area $A'$ decreases monotonically. Hence there is a unique value of $q$ at which the area is $\pi - \alpha - \beta - \gamma$. At that value of $q$ the angle $\beta'$ must equal $\beta$ by the Gauss–Bonnet theorem. $\qquad \square$

FACT 4. *If $\Delta = pqr$ is a triangle in hyperbolic space, and if $x$ is a point of the side $pq$, then there is a point $y \in pr \cup qr$ such that the hyperbolic distance $d(x, y)$ is less than $\ln(1 + \sqrt{2})$; that is, triangles in hyperbolic space are uniformly thin.*

PROOF. We need two observations. First, if $P$ and $Q$ are two vertical geodesics in the half-space model $H$ for hyperbolic space, and if a point $p$ moves monotonically downward along $P$, then the distance $d(p, Q)$ increases monotonically to infinity. See Figure 23. Second, if $p$ and $q$ are two points on the same boundary-orthogonal semicircle (geodesic) in $H$, say on the unit circle with coordinates $p = (\cos(\phi), \sin(\phi))$ and $q = (\cos(\theta), \sin(\theta))$ with $\theta > \phi$, then the hyperbolic distance between the two is given by the formula

$$d(p, q) = \int_{\phi}^{\theta} \frac{d\psi}{\sin(\psi)} = \ln \frac{\sin(\psi)}{1 + \cos(\psi)} \Big|_{\phi}^{\theta}.$$

See Figure 24. Actually, the radius of the semicircle is irrelevant because scaling is a hyperbolic isometry. Only the beginning and ending angles are important.

We are now ready for the proof that triangles are thin. Let $\Delta = pqr$ denote a triangle in the hyperbolic plane. We view $\Delta$ in the half-space model of the

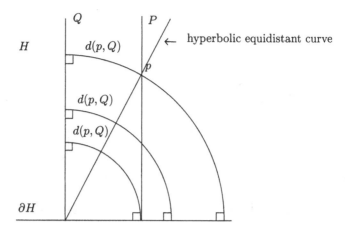

**Figure 23.** The monotonicity of $d(p, Q)$.

hyperbolic plane. We may assume that the side $pq$ lies in the unit circle with $p$ to the left of $q$, and we may assume that the side $pr$ is vertical with $r$ above $p$. We assume a point $x \in pq$ given. See Figure 25. We want to find an upper bound for the distance $d(x, pr \cup qr)$. The following operations simply expand the triangle $\Delta$ and hence increase the distance that we want to bound above. First we may move $r$ upward until it moves to $\infty$. We may then slide $p$ leftward along the unit circle until it meets infinity at $p' = -1$. We may then slide $q$ rightward along the unit circle until it meets infinity at $q' = 1$. We now have an ideal triangle $p'q'\infty$ with $x \in p'q'$. See Figure 26. The pair of sides $p'q'$ and $p'\infty$ are congruent as a pair to a pair of vertical geodesics (simply move $p'$ to $\infty$ by an isometry of $H$). Hence as we move $x$ toward $q'$, the distance $d(x, p'\infty)$ increases monotonically. Similarly, as we move $x$ toward $p'$, the distance $d(x, q'\infty)$ increases monotonically. We conclude that the maximum distance to $p'\infty \cup q'\infty$ is realized when $x$ is at the topmost point of the unit circle. The distances to the two vertical geodesics $p'\infty$ and $q'\infty$ are then equal and the shortest path is realized by a boundary-orthogonal semicircle that passes through $x$ and meets, say, $p'\infty$ orthogonally (if it did not meet orthogonally, then a shortcut near the vertical geodesic would reduce the length of the path). It is clear from the geometry that this shortest

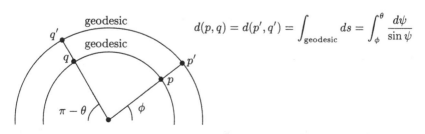

$$d(p, q) = d(p', q') = \int_{\text{geodesic}} ds = \int_{\phi}^{\theta} \frac{d\psi}{\sin \psi}$$

**Figure 24.** The formula for $d(p, q)$.

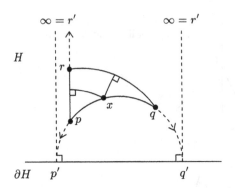

**Figure 25.** Triangles are thin.

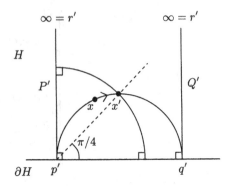

**Figure 26.** The ideal triangle $p'q'\infty$.

path travels through the angle interval $[\pi/4, \pi/2]$ in going from $x$ to the vertical geodesic $p'\infty$. Hence, by our calculation above, the distance between the point and the opposite sides is

$$\ln\frac{\sin(\pi/2)}{1+\cos(\pi/2)} - \ln\frac{\sin(\pi/4)}{1+\cos(\pi/4)} = \ln\left(1+\sqrt{2}\right) = 0.88\ldots.$$

We conclude that triangles are uniformly thin, as claimed.        □

FACT 5. *For a circular disk in the hyperbolic plane, the ratio of area to circumference is less than 1 and approaches 1 as the radius approaches infinity. That is, almost the entire area of the disk lies very close to the circular edge of the disk. Both area and circumference are exponential functions of hyperbolic radius.*

PROOF. We do our calculations in the disk model $I$ of the hyperbolic plane. The Riemannian metric is, as we recall,

$$ds_I^2 = 4(dx_1^2 + \cdots + dx_n^2)/(1 - x_1^2 - \cdots x_n^2)^2.$$

We are considering the case $n = 2$. Using polar coordinates (see Section 6) we can easily compute the distance element along a radial arc, namely

$$ds = 2\frac{dr}{1-r^2},$$

while the area element is

$$dA = \frac{4}{(1-r^2)^2} r\, dr\, d\theta.$$

We fix a Euclidean radius $R$ with associated circular disk centered at the origin in $I$ and calculate the hyperbolic radius $\rho$, area $A$, and circumference $C$ (see

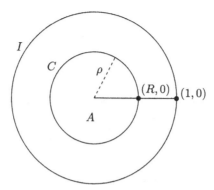

**Figure 27.** The hyperbolic radius $\rho$, the area $A$, and the circumference $C$.

Figure 27):

$$\rho = \int_0^R 2\,\frac{dr}{1-r^2} = \ln\frac{1+R}{1-R};$$

$$A = \int_{\theta=0}^{2\pi}\int_{r=0}^R \frac{4}{(1-r^2)^2}r\,dr\,d\theta = \frac{4\pi R^2}{1-R^2};$$

$$C = \int_{\theta=0}^{2\pi}\frac{2R}{1-R^2}\,d\theta = \frac{4\pi R}{1-R^2}.$$

Therefore

$$R = \frac{e^\rho-1}{e^\rho+1} = \frac{\cosh\rho-1}{\sinh\rho};$$

$$A = 2\pi(\cosh\rho-1) = 2\pi\left(\frac{\rho^2}{2!}+\frac{\rho^4}{4!}+\cdots\right) \approx \pi\rho^2 \text{ for small } \rho;$$

$$C = 2\pi\sinh\rho = 2\pi\left(\rho+\frac{\rho^3}{3!}+\frac{\rho^5}{5!}+\cdots\right) \approx 2\pi\rho \text{ for small } \rho.$$

Note that the formulas are approximately the Euclidean formulas for small $\rho$. This is apparent in the half-space model if one works near a point at unit Euclidean distance above the bounding plane; for at such a point the Euclidean and hyperbolic metrics coincide, both for areas and lengths. □

FACT 6. *In the half-space model $H$ of hyperbolic space, if $S$ is a sphere centered at a point at infinity $x \in \partial H$, then inversion in the sphere $S$ induces a hyperbolic isometry of $H$ that interchanges the inside and outside of $S$ in $H$.*

PROOF. Consider a Euclidean sphere $S$ centered at a point $p$ of the bounding plane at infinity. Let $x$ be an arbitrary point of $H$, and let $M$ be the Euclidean straight line through $p$ and $x$. There is a unique point $x' \in M \cap H$ on the opposite side of $S$ such that the two Euclidean straight line segments $x(S\cap M)$ and $x'(S\cap M)$ have the same *hyperbolic* length. See Figure 28. The points $x$ and $x'$ are said to be *mirror images* of one another with respect to $S$. We claim that

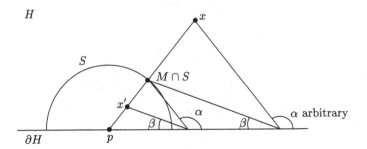

**Figure 28.** Inversion in $S$.

the map of $H$ that interchanges all of the inverse pairs $x$ and $x'$ is a hyperbolic isometry. We call this map *inversion in $S$*.

Note that all such spheres $S$ are congruent via hyperbolic isometries that are Euclidean similarities. Inversion is clearly invariant under such isometries. We shall make use of this fact both in giving formulas for inversion and in proving that inversion is a hyperbolic isometry.

Though our proof will make no use of formulas, we nevertheless describe inversion in $S$ by means of a formula. We lose no generality in assuming that $S$ is centered at the origin of Euclidean space. If $S$ has radius $r$ and if $x$ has length $t$, then multiplication of $H$ by the positive constant $r/t$ is a hyperbolic isometry that takes $M$ onto itself and takes $x$ to the point $M \cap S$. A second multiplication by $r/t$ takes the Euclidean segment $x(M \cap S)$ to a segment of the same hyperbolic length on the opposite side of $S$, hence takes $M \cap S$ to $x'$. That is, $x' = (r/t)^2 x$.

We now prove that inversion is a hyperbolic isometry. For that purpose we consider the hemisphere model $J$ for hyperbolic space. Consider the $n$-dimensional plane $P = \{x \in \mathbb{R}^{n+1} : x_1 = 0\}$ through the origin of $\mathbb{R}^{n+1}$ that is parallel to the half-space model $H = \{x \in \mathbb{R}^{n+1} : x_1 = 1\}$ of hyperbolic space. See Figure 29. This plane intersects the hemisphere model $J$ in one half of a sphere of dimension $n - 1$, which we denote by $S'$. The entire model $J$ is filled by circular segments that begin at the point $(-1, 0, \ldots, 0)$, end at the point $(1, 0, \ldots, 0)$, and intersect $S'$ at right angles. The hyperbolic metric $ds_J^2$ is clearly symmetric with respect to the plane $P$ and its intersection $S'$ with $J$. Euclidean reflection in that plane therefore induces a hyperbolic isometry of $J$ that takes a point on any of our circular segments to the point on the same circular segment but on the opposite side of $S'$. The symmetry of the hyperbolic metric clearly implies that the hyperbolic length of the two corresponding circular segments joining the point and its image to $S'$ have the same hyperbolic length.

Now map $J$ to $H$ by stereographic projection. Then $S'$ goes to one of our admissible spheres $S \cap H$ and our circular segments go to the family of lines $M$ through the origin. We see therefore that our hyperbolic reflection isometry

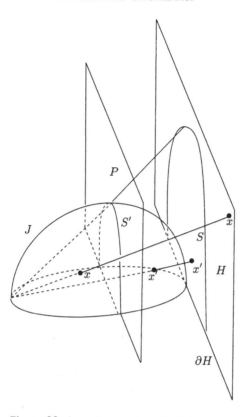

**Figure 29.** Inversion is a hyperbolic isometry.

of $J$ goes precisely to our inversion of $H$ in the sphere $S$. This completes the proof.                                                                                    □

## 14. The Sixth Model

We now turn to yet another model of the hyperbolic plane. This sixth model is only a combinatorial approximation to the half-space model, rather than a model in the sense of the other five. Consider the (infinite) family of "squares" sitting in the half-plane model, part of which is shown in Figure 30. This family is the image of the unit square, with vertical and horizontal sides and whose lower left corner is at $(0, 1)$, under the maps $p \mapsto 2^j\big(p + (k, 0)\big)$ with $(j, k) \in \mathbb{Z}^2$. Since horizontal translation and homotheties are hyperbolic isometries in $H$, each "square" is isometric to every other square. (We've called them squares even though in the hyperbolic metric they bear no resemblance to squares.)

Moving around in this family of squares is essentially like moving around in the hyperbolic plane. The advantage of the squares is that you can see combinatorially many of the aspects of hyperbolic space.

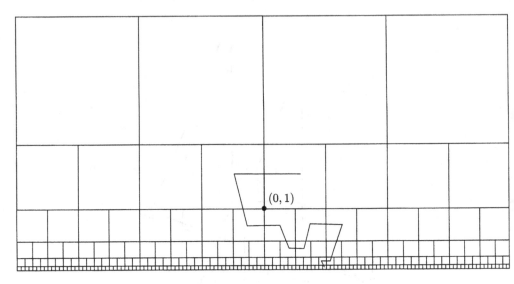

**Figure 30.** The sixth model and a random walk in the dual graph.

For example, note that a random walk on the dual graph will tend almost surely to infinity: from inside any square, the probability of exiting downwards is twice as great as the probability of exiting upwards.

Let $p$ and $q$ be vertices of the dual graph. Then one geodesic from $p$ to $q$ is gotten by taking a a path as in Figure 31, which rises initially straight upwards, goes horizontally a length at most 5, and then descends to $q$. More generally, let $\gamma$ be a geodesic from $p$ to $q$. Then there exists a geodesic $\delta$ from $p$ to $q$ that rises initially straight upwards, goes horizontally a length at most 5, and then descends to $q$ in such a way that the distance from any vertex of $\gamma$ is at most one from some vertex of $\delta$, and vice versa.

Another aspect of the hyperbolic plane that can be illustrated in this model is the "thin triangles" property. Given that we understand what geodesics look like from the previous paragraph, we first consider only a triangle with geodesic sides as in Figure 32.

The combinatorial lengths of the bottom two horizontal arcs are at most 5. Since the combinatorial length divides by approximately 2 as you ascend one level up, it follows that the combinatorial vertical distance from the middle horizontal arc to the top horizontal arc is at most 3. Hence it follows that every point on one side of the triangle is within distance at most 8 of the union of the two opposite sides of the triangle. Thus triangles in this model are said to be 8-*thin*. (In hyperbolic space, we saw that triangles are $\log(1 + \sqrt{2})$-thin in this sense.)

A consequence of the thin triangles property in a metric space is the exponential divergence of geodesics. Consider once again the half-space model $H$. Recall that a hyperbolic sphere (the set of points at a fixed distance from a point) is in fact also a Euclidean sphere.

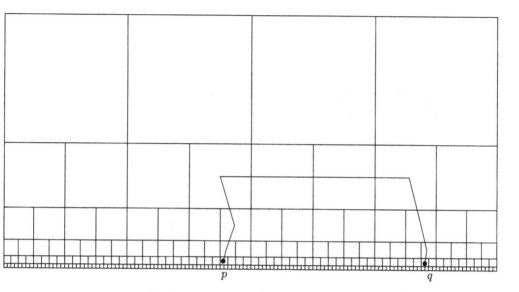

**Figure 31.** A geodesic in the dual graph.

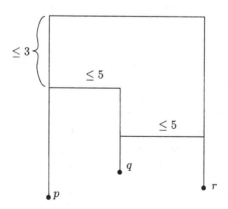

**Figure 32.** A triangle with geodesic sides.

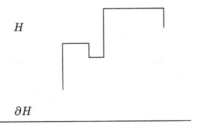

$H$

$\partial H$

**Figure 33.** A taxicab path. Integrate with respect to $ds = \sqrt{dx^2 + dy^2}/y$ to find the taxicab distance.

As in the proof of Fact 5 in Section 13, the area of a disk of radius $r_h$ is

$$A_h = 2\pi(\cosh(r_h) - 1),$$

whereas the length of the boundary of a disk of radius $r_h$ is $2\pi\sinh(r_h)$. For large $r_h$, these are both quite close to $\pi e^{r_h}$, so in particular we see that the circumference is exponential in the radius. This phenomenon will be known as the *exponential explosion*, and is true in any metric space satisfying the thin triangles condition.

Before we go on, we leave the reader with two exercises.

EXERCISE 1. Take a "taxicab" metric on $H^2$ in which the allowed paths are polygonal paths that have horizontal or vertical edges. See Figure 33. Analyze the geodesics in this new metric, and prove the thin triangles property.

EXERCISE 2. Generalize the previous exercise to $H^3$: let the allowed paths be polygonal paths that are vertical (in the $z$-direction) or horizontal (lie parallel to the $xy$-plane). Define the length of a horizontal line segment to be $\max\{\Delta x, \Delta y\}/z$.

## 15. Why Study Hyperbolic Geometry?

Hyperbolic geometry and its geometric insights have application in diverse areas of mathematics. The next three sections informally introduce some of those applications. The material is intended to be skimmed, since the reader may be unfamiliar with some of the prerequisite background material.

Hyperbolic geometry arises in three main areas:

(i) Complex variables and conformal mappings. In fact, work on automorphic functions was Poincaré's original motivation for defining hyperbolic space; see the quotation on page 63.

(ii) Topology (of three-manifolds in particular). More on this later regarding Thurston's surprising geometrization conjecture.

(iii) Group theory, in particular combinatorial group theory à la Gromov.

Historically, hyperbolic geometry lies at the center of a triangle around which revolve these three topics. See Figure 34. By using hard theorems in one domain and hard connections between domains, one can prove surprising results.

One such example is the Mostow rigidity theorem [1973], which we quote here in its simplest, lowest-dimensional form, for oriented, connected, compact three-dimensional manifolds. To state the theorem we need a definition: a *hyperbolic structure* on an $n$-dimensional manifold $M$ is a Riemannian metric on $M$ such that the universal covering space of $M$ is isometric to hyperbolic space $H^n$.

THEOREM 15.1. *Suppose the three-manifolds $M_1$ and $M_2$ having hyperbolic structures are homotopy equivalent. Then $M_1$ and $M_2$ are in fact isometric.*

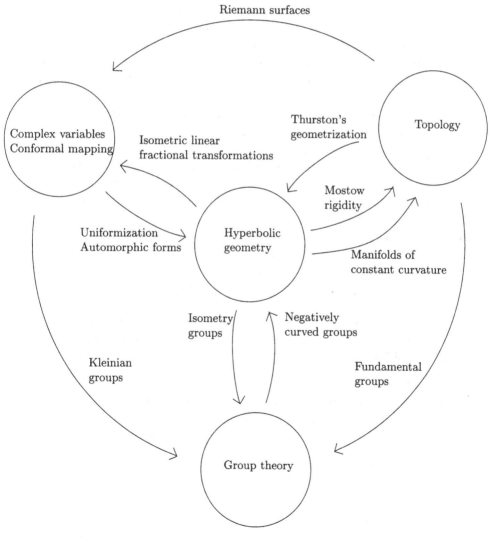

**Figure 34.** Connections between hyperbolic geometry and the three areas.

Here is a very abbreviated outline of the proof. There is an approximate metric correspondence between the universal covering space of $M_1$, as realized by hyperbolic space $H^3$, and the fundamental group $\pi_1(M_1)$, as realized geometrically by its Cayley group graph. Likewise the universal cover of $M_2$ and the Cayley graph of $\pi_1(M_2)$ are approximately isometric. The isomorphism between the fundamental groups $\pi_1(M_1)$ and $\pi_1(M_2)$ relates the two Cayley graphs metrically, and consequently gives an approximate geometric mapping from $H^3$, viewed as one universal cover, back to $H^3$ itself, viewed as the other universal cover. This mapping relates the actions of the two groups on their universal covers with enough precision that it matches their actions on the two-sphere at infinity homeomorphically. It remains to prove that this correspondence at infinity is not only homeomorphic but exactly conformal; the proof uses the theory of quasiconformal mappings and ergodicity of the group actions on the two-sphere at infinity. When the most general theorem is proved, the details fill a book.

A corollary to this result is that the hyperbolic structure on a three-manifold is a *topological invariant*. Consequently, so also are hyperbolic volume, lengths of geodesics, etc.

The proof of Mostow Rigidity suggests that many important properties of hyperbolic geometry are retained by spaces that only *approximate* hyperbolic geometry. The case in point involves the Cayley graphs of the fundamental groups, which are only one-dimensional and are not even manifolds, yet give a good approximation to three-dimensional hyperbolic geometry in the large. We shall return to this point when we describe Gromov's word-hyperbolic groups. But first we make precise the notion of spaces that are metrically comparable in the large, or *quasi-isometric*. We start by recalling some definitions.

A *group action* of a group $G$ on a space $X$ is a map $\alpha : G \times X \to X$, denoted $\alpha(g, x) = g(x)$, such that $1(x) = x$ for all $x \in X$, and $(g_1 g_2)(x) = g_1(g_2(x))$ for all $g_1, g_2 \in G$ and $x \in X$. In other words, $\alpha$ is a homomorphism from $G$ into $\mathrm{Homeo}(X)$.

A *geometry* is a path metric space in which metric balls are compact.

A *geometric action* of a group $G$ on a geometry $X$ is a group action that satisfies the following conditions:

(i) $G$ acts by isometries of $X$.
(ii) The action is properly discontinuous; that is, for every compact set $Y \subset X$, the set $\{g \in G : g(Y) \cap Y \neq \varnothing\}$ has finite cardinality.
(iii) The quotient $X/G = \{xG : x \in X\}$ is compact in the quotient topology.

Two geometries $X_1, X_2$ are *quasi-isometric* if there exist (not necessarily continuous) functions $R : X_1 \to X_2$, $S : X_2 \to X_1$ and a positive real number $M$ such that

(i) $S \circ R : X_1 \to X_1$ and $R \circ S : X_2 \to X_2$ are within $M$ of the identities.
(ii) For all $x_1, y_1 \in X_1$, $d(R(x_1), R(y_1)) \leq M d(x_1, y_1) + M$ and likewise for $X_2$.

THEOREM 15.2 (QUASI-ISOMETRY THEOREM). *If a group G acts geometrically on geometries $X_1$ and $X_2$, then $X_1$ and $X_2$ are quasi-isometric.*

Here are some exercises to challenge your understanding of these concepts.

EXERCISE 3. Let $G = \mathbb{Z}^2$, let $X_1$ be the Cayley graph of $G$ with standard generators, and let $X_2 = \mathbb{R}^2$. Show that $X_1$ and $X_2$ are quasi-isometric.

EXERCISE 4. If $G$ acts geometrically on any geometry, then $G$ is finitely generated.

EXERCISE 5. (Harder) If $G$ acts geometrically on any simply connected geometry, then $G$ is finitely presented.

EXERCISE 6. (Harder) If $G$ acts geometrically on any $n$-connected geometry, then $G$ has a $K(G, 1)$ with finite $(n + 1)$-skeleton.

(For proofs of 2, 3, and 4, see [Cannon 1991].)

**The Gromov–Rips thin triangles condition.** The most important *approximate* metric property of hyperbolic geometry is the *thin triangles condition:* let $\delta$ be a nonnegative constant; geodesic triangles in a path-metric space $X$ are $\delta$-*thin* if each point on each side of each geodesic triangle in $X$ lies within $\delta$ of the union of the other two sides of the triangle. A group is *negatively curved* (in the large) or *word-hyperbolic* if it is finitely generated, and with respect to some finite generating set, there is a nonnegative constant $\delta$ such that all of the geodesic triangles of the Cayley graph of the group are $\delta$-thin. Gromov has outlined a beautiful theory of negatively curved groups, and many mathematicians have helped to work out the details.

Many of the results are suggested directly by the classical case of hyperbolic geometry and the Kleinian groups which act geometrically on hyperbolic geometry.

We shall outline only one aspect of the Gromov program, namely the *space at infinity*.

**The space at infinity.** We have already noted that for each of our models $H, I, J, K, L$, there is a natural space at infinity: in the model $I$, for example, it is the unit $(n - 1)$-sphere that bounds $I$. This space at infinity can be seen from within the models themselves, as we indicated in the outline of Mostow's proof and in more detail now explain.

To each point "at infinity", there is a family of geodesic rays within the model that "meet" at the given point at infinity in a well-defined sense. Namely, define a *point at infinity* as an equivalence class of geodesic *rays*, any two being equivalent if they are asymptotically near one another (remain within a bounded distance of one another). See Figure 35. We let $S_\infty$ denote the set of such equivalence classes and call it the *space at infinity*.

We can define an intrinsic topology on the space at infinity as follows: given a single geodesic ray, the orthogonal complement at a point on the ray determines

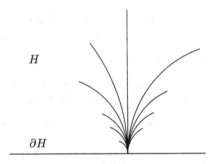

**Figure 35.** Geodesics with a common endpoint at infinity.

a hyperplane that bounds two hyperbolic half-spaces of hyperbolic space. One of these two half-spaces, the one containing the terminal subray of our ray, cuts off a disk on the sphere at infinity, and determines thereby a basic or fundamental neighborhood of the endpoint of the geodesic ray. See Figure 36. It is easy to see that this topology is invariant under hyperbolic isometries, and that the group of isometries acts as homeomorphisms of $S_\infty$.

Gromov [1987] has shown that an analogous space at infinity can always be defined for a space where triangles are uniformly thin. Though his construction is not exactly analogous to what we have just described, it is nevertheless possible to obtain exactly the Gromov space by a construction that *is* exactly analogous to what we have described [Cannon 1991]. In particular, one may define geodesic rays and equivalent rays, also half-spaces and fundamental "disks" at infinity. See Figure 37.

A special property of the classical spaces at infinity is that hyperbolic isometries act on the space at infinity not only as homeomorphisms but also as conformal mappings. This can be seen from the conformal models simply by the fact that the isometries preserve spheres in the ambient space $\mathbb{R}^{n+1}$, and so preserve spheres on $S_\infty$. The same is true of Gromov boundaries only in a weak sense.

For a deeper understanding of negatively curved groups and the related word-hyperbolic group, the reader should turn to the expository articles [Alonso et al.

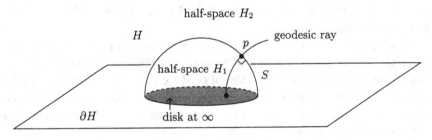

**Figure 36.** The disk at $\infty$ determined by a ray $p$ and a point on that ray. $S$ is the hyperbolic orthogonal complement to $p$.

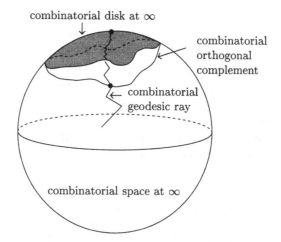

combinatorial disk at ∞

combinatorial orthogonal complement

combinatorial geodesic ray

combinatorial space at ∞

**Figure 37.** The combinatorial analogue.

1991; Cannon 1991; Coornaert, Delzant, and Papadopoulos 1990; Ghys and de la Harpe 1990] and Swenson's Ph.D. thesis [1993].

## 16. When Does a Manifold Have a Hyperbolic Structure?

Deciding when a manifold has a hyperbolic structure is a difficult problem. Much work has been done on this problem, and there are several hyperbolization conjectures and theorems. Let $M$ be a closed (compact, without boundary) three-manifold. If $M$ is hyperbolic, it is known that its fundamental group $\pi_1(M)$ satisfies these conditions:

- It is infinite.
- It does not contain a $\mathbb{Z} \oplus \mathbb{Z}$.
- It is not a free product.

*Thurston's hyperbolization conjecture* is that the converse is also true: these three conditions are also sufficient for $M$ to be hyperbolic. Thurston has proved this under some additional assumptions.

We now describe one of several programs attempting to prove the hyperbolization conjecture. This program involves at various stages all the connections of Figure 34; in fact one can trace the line of proof in a spiral fashion around the diagram in Figure 34. We start in the upper right corner.

The first step, Mosher's *weak hyperbolization conjecture* [Mosher 1995; Mosher and Oertel ≥ 1997] states that if $G = \pi_1(M)$ satisfies the above three conditions, then it has thin triangles (by which we mean that its Cayley graph $\Gamma(G)$ for some choice of generators has the thin triangles property). This brings us from topology into the domain of combinatorial group theory.

Note that in a group $G$ with the thin triangles property you can define the space at infinity $\partial G$, whose points are equivalence classes of geodesics (in $\Gamma(G)$) staying a bounded distance apart.

Assuming additionally that $\pi_1(M)$ has boundary homeomorphic to $S^2$, we attempt to equip this sphere with a conformal structure on which $\pi_1(M)$ acts uniformly quasiconformally. This would bring the problem into the domain of conformal mappings.

We would then apply a result of Sullivan and Tukia to conclude that the group acts conformally for another conformal structure, quasiconformally equivalent to this one.

Conformal self-maps of $S^2$ extend to hyperbolic isometries of $H^3$ (in the disk model $I$). This would then give us, by taking the quotient, a hyperbolic manifold (actually, an orbifold) $M'$ homotopy equivalent to $M$.

Gabai and collaborators [Gabai 1994a; Gabai 1994b; Gabai, Meyerhoff, and Thurston 1996] are extending Mostow rigidity to show that a three-manifold homotopy equivalent to a hyperbolic three-manifold $M'$ is in fact homeomorphic to $M'$.

So this would complete the program. Unfortunately many gaps remain to be bridged.

Our current focus is on the construction of a conformal structure assuming $\pi_1(M)$ has thin triangles and the space at infinity is homeomorphic to $S^2$. We have the following theorem (the converse of what we'd like to prove):

THEOREM 16.1. *Suppose a group $G$ acts geometrically on $H^3$. Then:*

1. *$G$ is finitely generated.*
2. *$\Gamma = \Gamma(G)$ (the Cayley graph for some choice of generators) has thin triangles.*
3. *$\partial\Gamma \cong S^2$.*

CONJECTURE 16.2. *The converse holds.*

Here is the intuition behind parts 2 and 3 in Theorem 16.1. The group $G$ acts geometrically on $\Gamma$ and on $H^3$. By the quasi-isometry theorem, $H^3$ and $\Gamma$ are quasi-isometric.

Consequently, the image in $H^3$ of a geodesic in $\Gamma$ looks "in the large" like a geodesic with a *linear* factor of inefficiency. To avoid exponential inefficiency, it must stay within a bounded distance of some genuine geodesic.

Any triangle in $\Gamma$ will map to a thin triangle in $H^3$, and hence is thin itself, which proves condition 2. Condition 3 is established similarly.

To understand the difficulty in proving the conjecture, we have to appreciate the difference between constant and variable negative sectional curvature.

Consider the following example, which illustrates a variable negative curvature space. In the space $K^3 = \{(x, y, z) : z > 0\}$, consider the paths that are piecewise vertical (in the $z$-direction) or horizontal (parallel to the $xy$-plane). Use the metric length element $|dz|/z$ for vertical paths, and the metric $\max\{|dx|/z^a, |dy|/z^b\}$ (where $a, b > 0$ are constants) for horizontal paths.

This metric is analogous to the Riemannian metric

$$ds^2 = \frac{dx^2}{z^{2a}} + \frac{dy^2}{z^{2b}} + \frac{dz^2}{z^2},$$

but the calculations are simpler. Note that the latter reduces to the hyperbolic metric when $a = b = 1$.

In a plane parallel to the $xz$-plane our metric is analogous to

$$\frac{dx^2}{z^{2a}} + \frac{dz^2}{z^2},$$

which, under the change of variables $X = ax$, $Z = z^a$, yields the metric

$$\frac{1}{a^2} \frac{dX^2 + dZ^2}{Z^2};$$

the latter is a scaled version of the hyperbolic metric. A similar formula holds for the planes parallel to the $yz$ plane. If $a \neq b$ then these two sectional curvatures are indeed different.

It is not hard to figure out what the geodesics in $K^3$ look like. A shortest (piecewise horizontal and vertical) curve joining two points $p_1 = (x_1, y_1, z_1)$ and $p_2 = (x_2, y_2, z_2)$ goes straight up from $p_1$ to some height $z_3$, then goes horizontally and straight in the plane $z = z_3$ until it is above $p_2$, and then goes straight down to $p_2$. Since the length of such a path is

$$\ell(z_3) = \log(z_3/z_1) + \log(z_3/z_2) + \max\{|x_1 - x_2|z_3^{-a}, |y_1 - y_2|z_3^{-b}\},$$

we can then find the optimal $z_3$ by differentiating and considering the various cases.

Consider a geodesic line of the form $p(t) = (x_1, y_1, z_1 e^{-t})$. The half space specified by that line and the point $p(0)$ turns out to be the box

$$B = \left\{(x, y, z) : |x - x_1| < \frac{2z_1^b}{b}, \ |y - y_1| < \frac{2z_1^a}{a}, \ 0 < z < z_1\right\}.$$

The footprint of this half-space on the space at infinity is the rectangle

$$\left\{(x, y) : |x - x_1| < 2\frac{z_1^b}{b}, \ |y - y_1| < 2\frac{z_1^a}{a}\right\},$$

whose aspect ratio is $az_1^{b-a}/b$. These aspect ratios are not bounded, so the half spaces do not induce any reasonable conformal structure at infinity.

Note that the isometries of $K^3$ include horizontal translations and maps of the form $(x, y, z) \to (v^a x, v^b y, vz)$. The latter map acts linearly on the space at infinity. However, for large $v$, the quasiconformal distortion is unbounded (when $a \neq b$).

## 17. How to Get Analytic Coordinates at Infinity?

The previous example suggests that the task of finding analytic coordinates on $S^2$ for which the group acts uniformly quasiconformally may be difficult. Among the uncountably many quasiconformality classes of conformal structures on a topological $S^2$, one must select (the unique) one on which the group acts uniformly quasiconformally.

In order to accomplish this task, one needs to work with whatever structure on the sphere is a priori provided by the group. Let $v_0$ be the vertex of $\Gamma$ corresponding to the identity of $G$. Fix some positive integer $n$. Consider the collection of all combinatorially defined half-spaces defined by any geodesic ray starting at $v_0$ and the vertex on the ray at distance $n$ from $v_0$ [Cannon 1991]. These half-spaces cut off combinatorial "disks" at infinity and thereby give a finite covering of $S^2$. In the appropriate conformal structure on $S^2$ (if it exists), the sets in this cover are approximately round [Cannon and Swenson $\geq$ 1997]. Hence we should think of this cover as providing a sort of "discrete conformal structure" on $S^2$.

The uniformization theorem for $S^2$ says that any conformal structure on $S^2$ is equivalent to the standard Riemann sphere. Hence, once a conformal structure is constructed, analytic coordinates exist. This suggests that one should look for discrete generalizations of uniformization theorems, and in particular, of the Riemann mapping theorem.

The Riemann mapping theorem is a theorem about conformal mappings, and conformality is usually defined in terms of analytic derivatives. In the absence of a priori analytic coordinates, any discrete Riemann mapping theorem cannot begin with a well-defined notion of analytic derivative. Fortunately, there are variational formulations of the Riemann mapping theorem that avoid the mention of derivatives. One is based on *extremal length*.

Consider a quadrilateral $Q$ in the plane $\mathbb{C}$. This is just a closed topological disk with four distinct points marked on the boundary. These marked points partition the boundary of the disk into four arcs, denoted $a, b, c, d$ in Figure 38. Consider metrics on $Q$ that are conformal to the metric that $Q$ inherits from the plane. Conformal changes of metric are determined by positive weight functions $m : Q \to (0, \infty)$ that one should view as point-by-point scalings of the Euclidean metric. With such a weight function $m$ one can define (weighted) lengths of paths $\gamma$ by

$$\ell_m(\gamma) = \int_\gamma m \, |dz|$$

and (weighted) total areas by

$$a_m = \int_Q m^2 \, dz \, d\bar{z}.$$

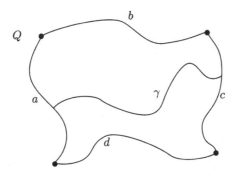

**Figure 38.** A quadrilateral $Q$.

Let $d_m$ be the distance in the weighted metric between the edges $a$ and $c$ of $Q$. It turns out that there is an essentially unique weight function $m_0$ that maximizes the ratio $d_m^2/a_m$, and $(Q, d_{m_0})$ is isometric with a rectangle. (Actually, $m_0$ is unique up to a positive scalar multiple and a.e. equivalence. Here we take $m_0$ to be the continuous representative in its a.e. equivalence class.) The maximal ratio $d_{m_0}^2/a_{m_0}$ is also called the extremal length from $a$ to $c$. It may be interpreted as the resistance to the flow of electricity between $a$ and $c$ if the quadrilateral is interpreted as a conducting metal plate.

This approach provides a uniformization theorem that does not resort to derivatives. It also has a discrete counterpart. (See Figure 39. For more information, see [Cannon 1994; Cannon, Floyd, and Parry 1994].)

A finite covering $C = \{C_j\}$ of a quadrilateral or annulus $Q'$ provides us with a discrete extremal length. In this discrete setting, a weight function is just an assignment of a nonnegative number $m(C_j)$ to each set $C_j$ in the covering. A length of a path $\gamma$ in $Q$ can be defined as just the sum of $m(C_j)$ over all $C_j \in C$ that intersect $\gamma$, and the area of $m$ is defined as the sum of $m(C_j)^2$ over all sets $C_j \in C$. We can then solve a discrete version of the extremal length problem on $Q'$, and use the solution to define an "approximate conformal structure".

This technique can be applied to find a conformal structure on $S^2 = \partial G$, if it exists: the half-spaces defined by $G$ as $n$ increases define a nested sequence of covers $C^n$ of $S^2$; we get a sequence of "finite" conformal structures that must converge, in the appropriate sense, to a genuine quasiconformal structure if one exists.

In this respect, we close with the following theorem.

THEOREM 17.1 (CANNON, FLOYD, PARRY). *There exists an invariant conformal structure on $S^2$ if and only if the sequence of covers $C^n$ satisfies the following: for every $x \in S^2$ and for every neighborhood $U$ of $x$, there is an annulus $Q$ whose closure lies in $U \setminus \{x\}$ and that separates $x$ from $S^2 \setminus U$, such that the discrete extremal lengths between the boundary components of $Q$ with respect to the sequence of covers $C^n$ are bounded away from $0$.*

# References

[Alonso et al. 1991]  J. M. Alonso et al., "Notes on word hyperbolic groups", pp. 3–63 in *Group theory from a geometrical viewpoint* (Trieste, 1990), edited by E. Ghys et al., World Scientific, River Edge, NJ, 1991.

[Benedetti and Petronio 1992]  R. Benedetti and C. Petronio, *Lectures on hyperbolic geometry*, Universitext, Springer, Berlin, 1992.

[Bolyai and Bolyai 1913]  F. Bolyai and J. Bolyai, *Geometrische Untersuchungen*, Teubner, Leipzig and Berlin, 1913. Edited by P. G. Stackel. Reprinted by Johnson Reprint Corp., New York and London, 1972. Historical and biographical materials.

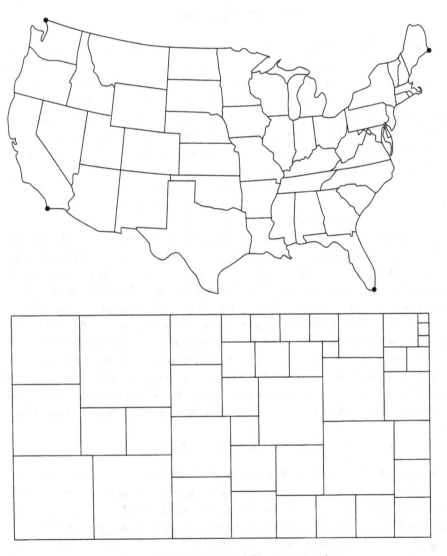

**Figure 39.** Combinatorial Riemann mapping.

[Cannon 1991] J. W. Cannon, "The theory of negatively curved spaces and groups", pp. 315–369 in *Ergodic theory, symbolic dynamics and hyperbolic spaces*, edited by T. Bedford et al., Oxford U. Press, Oxford, 1991.

[Cannon 1994] J. W. Cannon, "The combinatorial Riemann mapping theorem", *Acta Math.* **173** (1994), 155–234.

[Cannon et al. 1994] J. W. Cannon, W. J. Floyd, and W. R. Parry, "Squaring rectangles: the finite Riemann mapping theorem", pp. 133–212 in *The mathematical legacy of Wilhelm Magnus: groups, geometry and special functions* (Brooklyn, NY, 1992), edited by W. Abikoff et al., Contemporary Mathematics **169**, Amer. Math. Soc., Providence, 1994.

[Cannon and Swenson ≥ 1997] J. W. Cannon and E. L. Swenson, "Recognizing constant curvature groups in dimension 3". Preprint.

[Coornaert et al. 1990] M. Coornaert, T. Delzant, and A. Papadopoulos, *Géometrie et théorie des groupes: les groupes hyperboliques de Gromov*, Lecture Notes in Math. **1441**, Springer, Berlin, 1990.

[Euclid 1926] Euclid, *The Thirteen Books of Euclid's Elements*, second ed., Cambridge U. Press, Cambridge, 1926. Edited by T. L. Heath. Reprinted by Dover, New York, 1956.

[Gabai 1994a] D. Gabai, "Homotopy hyperbolic 3-manifolds are virtually hyperbolic", *J. Amer. Math. Soc.* **7** (1994), 193–198.

[Gabai 1994b] D. Gabai, "On the geometric and topological rigidity of hyperbolic 3-manifolds", *Bull. Amer. Math. Soc. (N.S.)* **31** (1994), 228–232.

[Gabai et al. 1996] D. Gabai, R. Meyerhoff, and N. Thurston, "Homotopy hyperbolic 3-manifolds are hyperbolic", 1996. Available at http://www.msri.org/MSRI-preprints/online/1996-058.

[Ghys and de la Harpe 1990] E. Ghys and P. de la Harpe, *Sur les groupes hyperboliques d'après Mikhael Gromov*, Birkhäuser, boston, 1990.

[Gromov 1987] M. L. Gromov, "Hyperbolic groups", pp. 75–263 in *Essays in group theory*, edited by S. Gersten, MSRI Publications **8**, Springer, New York, 1987. Perhaps the most influential recent paper in geometric group theory.

[Hilbert and Cohn-Vossen 1932] D. Hilbert and S. Cohn-Vossen, *Anschauliche Geometrie*, Grundlehren der math. Wissenschaften **37**, Springer, Berlin, 1932. Reprinted by Dover, New York, 1944. Translated by Paul Nemenyi as *Geometry and the Imagination*, Chelsea, New York, 1952. College-level exposition of rich ideas from low-dimensional geometry, with many figures.

[Iversen 1993] B. Iversen, *Hyperbolic geometry*, London Math. Soc. Student Texts **25**, Cambridge U. Press, Cambridge, 1993. Very clean algebraic approach to hyperbolic geometry.

[Klein 1928] F. Klein, *Vorlesungen über nicht-Euklidische Geometrie*, Grundlehren der math. Wissenschaften **26**, Springer, Berlin, 1928. Edited by W. Rosemann. Reprinted by Chelsea, New York, 1960. Mostly algebraic development of non-Euclidean geometry with respect to Klein and projective models. Beautiful figures. Elegant exposition.

[Kline 1972]  M. Kline, *Mathematical thought from ancient to modern times*, Oxford U. Press, New York, 1972. A well-written and fairly concise history of mathematics (1200 pages in one or three volumes). Full of interesting material.

[Lobachevskiĭ 1898]  N. I. Lobachevskiĭ, *Zwei geometrische Abhandlungen*, Bibliotheca math. Teubneriana **70**, Teubner, Leipzig, 1898. Reprinted by Johnson Reprint Corp., New York and London, 1972. Original papers.

[Mosher 1995]  L. Mosher, "Geometry of cubulated 3-manifolds", *Topology* **34** (1995), 789–814.

[Mosher and Oertel ≥ 1997]  L. Mosher and U. Oertel, "Spaces which are not negatively curved". Preprint.

[Poincaré 1908]   H. Poincaré, *Science et méthode*, E. Flammarion, Paris, 1908. Translated by F. Maitland as *Science and method*, T. Nelson and Sons, London, 1914. Reprinted by Dover, New York, 1952, and by Thoemmes Press, Bristol, 1996. One of Poincaré's several popular expositions of science. Still worth reading after almost 100 years.

[Ratcliffe 1994]  J. G. Ratcliffe, *Foundations of hyperbolic manifolds*, Graduate Texts in Math. **149**, Springer, New York, 1994. Fantastic bibliography, careful and unified exposition.

[Riemann 1854]  G. F. B. Riemann, "Über die Hypothesen, welche der Geometrie zu Grunde liegen", *Königliche Gesellschaft der Wissenschaften zu Göttingen* **13** (1854), 1–20. Also pp. 272–287 in *Gesammelte Mathematische Werke*, edited by R. Dedekind and H. Weber, 2nd ed., Teubner, Leipzig, 1902 (reprint: Dover, New York, 1953).

[Swenson 1993]  E. L. Swenson, *Negatively curved groups and related topics*, Ph.D. thesis, Brigham Young University, 1993.

[Thurston 1997]  W. P. Thurston, *Three-dimensional geometry and Topology*, Princeton U. Press, Princeton, 1997. Thurston reintroduced hyperbolic geometry to the topologist in the 1970s. This richly illustrated book has an accessible beginning and progresses to great and exciting heights.

[Weyl 1919]   H. Weyl, *Raum, Zeit, Materie: Vorlesungen über allgemeine Relativitätstheorie*, Springer, Berlin, 1919. Translated by H. L. Brose as *Space–Time–Matter*, Dover, New York, 1922; reprinted 1950. Weyl's exposition and development of relativity and gauge theory; begins at the beginning with motivation, philosophy, and elementary developments as well as advanced theory.

# Index

JAMES W. CANNON
DEPARTMENT OF MATHEMATICS
BRIGHAM YOUNG UNIVERSITY
PROVO, UT 84602-1001
cannon@math.byu.edu

WILLIAM J. FLOYD
DEPARTMENT OF MATHEMATICS
VIRGINIA POLYTECHNICAL INSTITUTE AND STATE UNIVERSITY
BLACKSBURG, VA 24061-0123
floyd@math.vt.edu

RICHARD KENYON
DÉPARTEMENT DE MATHÉMATIQUES
ÉCOLE NORMALE SUPÉRIEURE DE LYON
46, ALLÉE D'ITALIE, 69364 LYON
FRANCE
rkenyon@umpa.ens-lyon.fr

WALTER R. PARRY
DEPARTMENT OF MATHEMATICS
EASTERN MICHIGAN UNIVERSITY
YPSILANTI, MI 48197
mth_parry@emuvax.emich.edu

Flavors of Geometry
MSRI Publications
Volume **31**, 1997

# Complex Dynamics in Several Variables

## JOHN SMILLIE

### Notes by GREGERY T. BUZZARD

## CONTENTS

## 1. Motivation

The study of complex dynamics in several variables can be motivated in at least two natural ways. The first is by analogy with the fruitful study of complex dynamics in one variable. Since this latter subject is the subject of the parallel lectures by John Hubbard (the reader is referred to [Carleson and Gamelin 1993] for a good introduction to the subject), we focus here on the second source of motivation: the study of real dynamics.

A classical problem in the study of real dynamics is the $n$-body problem, which was studied by Poincaré. For instance, we can think of $n$ planets moving in space. For each planet, there are three coordinates giving the position and three coordinates giving the velocity, so that the state of the system is determined by a total of $6n$ real variables. The evolution of the system is governed by Newton's laws, which can be expressed as a first order ordinary differential equation. In fact, the state of the system at any time determines the entire future and past evolution of the system.

To make this a bit more precise, set $k = 6n$. Then the behavior of the $n$ planets is modeled by a differential equation

$$(\dot{x}_1, \ldots, \dot{x}_k) = F(x_1, \ldots, x_k)$$

for some $F : \mathbb{R}^k \to \mathbb{R}^k$. Here $\dot{x}$ denotes the derivative of $x$ with respect to $t$.

From the elementary theory of ordinary differential equations, we know that this system has a unique solution $t \mapsto \varphi_t(x_1, \ldots, x_k)$ satisfying $\dot{\varphi} = F(\varphi)$ and $\varphi_0(x_1, \ldots, x_k) = (x_1, \ldots, x_k)$.

For purposes of studying dynamics, we would like to be able to say something about the evolution of this system over time, given some initial data. That is, given $p \in \mathbb{R}^k$, we would like to be able to say something about $\varphi_t(p)$ as $t$ varies. For instance, a typical question might be the following.

QUESTION 1.1. *For given initial positions and velocities, do the planets have bounded orbits for all (positive) time? That is, given $p = (x_1, \ldots, x_k)$, is the set $\{\varphi_t(p) : t \geq 0\}$ bounded?*

This question, in fact, particularly interested Poincaré. Unfortunately, the usual answer to such a question is "I don't know." Nevertheless, it is possible to say something useful about related questions, at least in some settings. For instance, one related problem is the following.

PROBLEM 1.2. Say something interesting about the set of initial conditions for which the planets have bounded forward orbits. That is, describe the set

$$K^+ := \{p \in \mathbb{R}^k : \{\varphi_t(p) : t \geq 0\} \text{ is bounded}\}.$$

Although this question is less precise and gives less specific information than the original, an answer to it can still tell us quite a bit about the behavior of the system.

## 2. Iteration of Maps

In the preceding discussion, we have been taking the approach of fixing a point $p \in \mathbb{R}^k$ and following the evolution of the system over time starting from this point. An alternative approach is to think of all possible starting points evolving simultaneously, then taking a snapshot of the result at some particular instant in time.

To make this more precise, assume that the solution $\varphi_t(p)$ exists for all time $t$ and all $p \in \mathbb{R}^k$. In this case, for fixed $t$, the map $\varphi_t : \mathbb{R}^k \to \mathbb{R}^k$ is a diffeomorphism of $\mathbb{R}^k$ and satisfies the group property

$$\varphi_{s+t} = \varphi_s \circ \varphi_t$$

for any $s$ and $t$. The family of diffeomorphisms $(\varphi_t)_{t \in \mathbb{R}}$ is called the *flow* of the differential equation.

In order to make our study more tractable, we make two simplifications.

**Simplification 1.** Choose some number $\alpha > 0$, called the *period*, and define $f = \varphi_\alpha$. Then $f$ is a diffeomorphism of $\mathbb{R}^k$ and, given $p \in \mathbb{R}^k$, the group property of $\varphi$ implies that

$$\varphi_{n\alpha}(p) = \varphi_\alpha \circ \cdots \circ \varphi_\alpha(p) = f^n(p).$$

That is, studying the behavior of $f$ under iteration is equivalent to studying the behavior of $\varphi$ at regularly spaced time intervals. By concentrating on $f$ we ignore those aspects of the behavior of the continuous flow $\varphi$ that occur at time scales less than $\alpha$.

**Simplification 2.** Set $k = 2$. Although this simplification means that we can no longer directly relate our model to the original physical problem, the ideas and techniques involved in studying such a simpler model are still rich enough to shed some light on the more realistic cases. In fact, there are interesting questions in celestial mechanics which reduce to questions about two-dimensional diffeomorphisms, but here we are focusing on the mathematical model rather than on the physical system.

We also introduce some notation. Given $p \in \mathbb{R}^2$, let $\mathcal{O}^+(p)$, $\mathcal{O}^-(p)$, and $\mathcal{O}(p)$ be respectively the *forward orbit*, *backward orbit*, and *full orbit* of $p$ under $f$. In symbols,

$$\mathcal{O}^+(p) := \{f^n(p) : n \geq 0\},$$
$$\mathcal{O}^-(p) := \{f^n(p) : n \leq 0\},$$
$$\mathcal{O}(p) := \{f^n(p) : n \in \mathbb{Z}\}.$$

Problem 1.2 then becomes the following.

PROBLEM 2.1. Given a diffeomorphism $f : \mathbb{R}^2 \to \mathbb{R}^2$, describe the sets

$$K^+ := \{p \in \mathbb{R}^2 : \mathcal{O}^+(p) \text{ is bounded}\},$$
$$K^- := \{p \in \mathbb{R}^2 : \mathcal{O}^-(p) \text{ is bounded}\},$$
$$K := \{p \in \mathbb{R}^2 : \mathcal{O}(p) \text{ is bounded}\},$$

For future reference, note that $K = K^+ \cap K^-$.

## 3. Regular Versus Chaotic Behavior

For the moment, we will make no attempt to define rigorously what we mean by regular or chaotic. Intuitively, one should think of regular behavior as being very predictable and as relatively insensitive to small changes in the system or initial conditions. On the other hand, chaotic behavior is in some sense random and can change drastically with only slight changes in the system or initial conditions. Here is a relevant quote from Poincaré on chaotic behavior:

> A very small cause, which escapes us, determines a considerable effect which we cannot ignore, and we say that this effect is due to chance.

We next give some examples to illustrate both kinds of behavior, starting with regular behavior. First we make some definitions.

A point $p \in \mathbb{R}^2$ is a *periodic point* if $f^n(p) = p$ for some $n \geq 1$. The smallest such $n$ is the *period* of $p$. A periodic point $p$ is *hyperbolic* if $(Df^n)(p)$ has no eigenvalues on the unit circle. (Here $Df^n$ represents the derivative of $f^n$ at $p$, a linear map $\mathbb{R}^2 \to \mathbb{R}^2$.) If $p$ is a hyperbolic periodic point and both eigenvalues are inside the unit circle, $p$ is called a *sink* or *attracting periodic point*.

Let $d$ denote Euclidean distance in $\mathbb{R}^2$. If $p$ is a hyperbolic periodic point, the set

$$W^s(p) = \{q \in \mathbb{R}^2 : d(f^n q, f^n p) \to 0 \text{ as } n \to \infty\} \qquad (3.1)$$

is called the *stable manifold* of $p$: it is the set of points whose forward images become increasingly closer to the corresponding images of $p$. Dually, the *unstable manifold* of $p$ is made up of those points whose *backward* images approach those of $p$:

$$W^u(p) = \{q \in \mathbb{R}^2 : d(f^{-n} q, f^{-n} p) \to 0 \text{ as } n \to \infty\}. \qquad (3.2)$$

(The notation $f^{-n}$ represents the $n$-th iterate of $f^{-1}$, which is well defined since we are assuming that $f$ is a diffeomorphism.) When $p$ is a sink, the stable manifold $W^s(p)$ is also called the *attraction basin* of $p$.

FACT. *When $p$ is a sink, $W^s(p)$ is an open set containing $p$, and $W^u(p)$ is empty.*

A sink gives a prime example of regular behavior. Starting with any point $q$ in the basin of attraction of a sink $p$, the forward orbit of $q$ is asymptotic to the (periodic) orbit of $p$. Since the basin is open, this will also be true for any point $q'$ near enough to $q$. Hence we see the characteristics of predictability and stability mentioned in relation to regular behavior.

For an example of chaotic behavior, we turn to a differential equation studied by Cartwright and Littlewood in 1940, and given by

$$\ddot{y} - k(1 - y^2)\dot{y} + y = b \cos t.$$

Introducing the variable $x = \dot{y}$, we can write this as a first-order system

$$\dot{y} = x, \qquad \dot{x} = g(x, y, t),$$

where $g$ is a function satisfying $g(x, y, t + 2\pi) = g(x, y, t)$. This system has a solution $\varphi_t$ as before with $\varphi_t : \mathbb{R}^2 \to \mathbb{R}^2$ a diffeomorphism. Although the full group property does not hold for $\varphi$ since $g$ depends on $t$, we still have $\varphi_{s+t} = \varphi_s \circ \varphi_t$ whenever $s = 2\pi n$ and $t = 2\pi m$ for integers $n$ and $m$. Hence we can again study the behavior of this system by studying the iterates of the diffeomorphism $f = \varphi_{2\pi}$.

Rather than study this system itself, we follow the historical development of the subject and turn to a more easily understood example of chaotic behavior which was motivated by this system of Cartwright and Littlewood: the Smale horseshoe.

## 4. The Horseshoe Map and Symbolic Dynamics

The horseshoe map was first conceived by Steve Smale as a way of capturing many of the features of the Cartwright–Littlewood map in a system that is easily understood.

For our purposes, the horseshoe map, $h$ is defined first on a square $B$ in the plane with sides parallel to the axes. First we apply a linear map that stretches the square in the $x$-direction and contracts it in the $y$-direction. Then we take the right edge of the resulting rectangle and bend it around to form a horseshoe shape. The map $h$ is then defined on $B$ by placing this horseshoe over the original square $B$ so that $B \cap h(B)$ consists of two horizontal strips in $B$. See Figure 1.

We can extend $h$ to a diffeomorphism of $\mathbb{R}^2$ in many ways. We do it here as follows. First partition $\mathbb{R}^2 \setminus B$ into four regions by using the lines $y = x$ and $y = -x$ as boundaries. Denote the union of the two regions above and below $B$ by $B^+$ and the union of the two regions to the left and right of $B$ by $B^-$, as in Figure 2. Then we can extend $h$ to a diffeomorphism of $\mathbb{R}^2$ in such a way that $h(B^-) \subseteq B^-$. In this situation, points in $B^+$ can be mapped to any of the three regions $B^+$, $B$, or $B^-$, points in $B$ can be mapped to either $B$ or $B^-$, and points

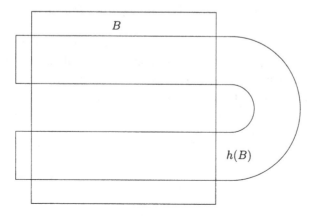

**Figure 1.** The image of $B$ under the horseshoe map $h$.

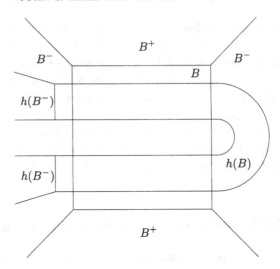

**Figure 2.** The sets $B$, $B^+$, and $B^-$.

in $B^-$ must be mapped to $B^-$. Further, we require that points in $B^-$ go to $\infty$ under iteration, and we require analogous conditions on $f^{-1}$. Note in particular that points that leave $B$ do not return and that $K \subseteq B$.

It is not hard to see that, under these conditions,

$$K^- \cap B = B \cap hB \cap h^2 B \cap \cdots.$$

In fact, if we look at the image of the two strips $B \cap hB$ and intersect with $B$, the resulting set consists of four strips; each of the original two strips is subdivided into two smaller strips. Continuing this process, we see that $K^- \cap B$ is simply the set product of an interval and a Cantor set.

In fact, a simple argument shows that $h$ has a fixed point $p$ in the upper left corner of $B$, and that the unstable manifold of $p$ is dense in the set $K^- \cap B$ and the stable manfold of $p$ is dense in $K^+ \cap B$. The complicated structure of the stable and unstable manifolds plays an important role in the behavior of the horseshoe map.

We can describe the chaotic behavior of the horseshoe using *symbolic dynamics*. The idea of this procedure is to translate from the dynamics of $h$ restricted to $K$ into the dynamics of a shift map on bi-infinite sequences of symbols.

To do this, first label the two components of $B \cap hB$ with $H_0$ and $H_1$. Then, to each point $p \in K$, associate a bi-infinite sequence of 0's and 1's (that is, an element of $\{0, 1\}^{\mathbb{Z}}$) using the map

$$\psi : p \mapsto s = (\ldots, s_1, s_0, s_{-1}, \ldots),$$

where

$$s_j = \begin{cases} 0 & \text{if } h^j(p) \in H_0, \\ 1 & \text{if } h^j(p) \in H_1. \end{cases}$$

We can put a metric on the space of bi-infinite sequences of 0's and 1's by

$$d(s, s') = \sum_{j=-\infty}^{\infty} |s_j - s'_j| 2^{-|j|}.$$

It is not hard to show that the metric space thus obtained is compact and that the map $\psi$ given above produces a homeomorphism between $K$ and this space $\{0, 1\}^{\mathbb{Z}}$ of sequences. For bi-infinite sequences we have the natural concept of a *shift map*, which shifts all the entries of a sequence by one position. Formally, the *left shift map* on $\{0, 1\}^{\mathbb{Z}}$ is the map that associates to a sequence $s = (s_i)_{i \in \mathbb{Z}}$ the sequence $z = (z_i)_{i \in \mathbb{Z}}$ defined by $z_i = s_{i+1}$. The definition of $\psi(p)$ implies that if $\sigma$ is the left-shift map defined on bi-infinite sequences, then $\psi(h(p)) = \sigma(\psi(p))$.

Here are a couple of simple exercises that illustrate the power of using symbolic dynamics.

EXERCISE 4.1. Show that periodic points are dense in $K$. Hint: Periodic points correspond to periodic sequences.

EXERCISE 4.2. Show that there are periodic points of all periods.

# 5. Hénon Maps

The horseshoe was one motivating example for what are known as Axiom A diffeomorphisms [Bowen 1978]. The features that make the horseshoe easy to analyze dynamically are the uniform expansion in the horizontal direction and the uniform contraction in the vertical direction. This behavior is captured in the notion of hyperbolicity. We say that a diffeomorphism is *hyperbolic* over a set $X \subset \mathbb{R}^2$ if for each $x \in X$ there is a direction in which length is uniformly expanded and a direction in which length is uniformly contracted. These directions can depend on the point $x$, but the angle between them must be bounded away from zero. Hyperbolicity is the key ingredient in the definition of Axiom A. Like the horseshoe, Axiom A diffeomorphisms admit a symbolic description. Another important property of Axiom A diffeomorphisms is structural stability. This implies that small changes in the parameters do not change the symbolic description of the diffeomorphism.

Axiom A diffeomorphisms received a great deal of attention in the 1960s and 70s. Much current work focuses either on how Axiom A fails, as in the work of Newhouse, or on how some Axiom A ideas can be applied in new settings, as in the work of Benedicks and Carleson [1991] or Benedicks and Young [1993]. For more information and further references, [Ruelle 1989] provides a fairly gentle introduction, while [Palis and de Melo 1982; Shub 1978; Palis and Takens 1993] are more advanced. See also [Yoccoz 1995].

A model system for the study of non-Axiom A behavior that has received a great deal of attention is the so-called Hénon map. This is actually a family of

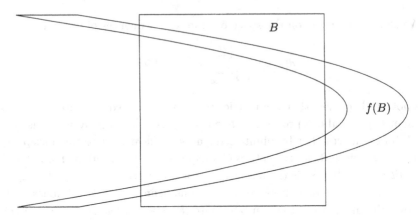

**Figure 3.** A square $B$ and its image $f(B)$ for some parameter values $a$ and $b$.

diffeomorphisms $f_{a,b} : \mathbb{R}^2 \to \mathbb{R}^2$ defined by

$$f_{a,b}(x,y) = (-x^2 + a - by,\ x)$$

for $b \neq 0$. These maps arise from a simplification of a simplification of a map describing turbulent fluid flow.

We can get some idea of the behavior of the map $f_{a,b}$ and the ways in which it relates to the horseshoe map by considering the image of a large box $B$ under $f_{a,b}$. For simplicity, we write $f$ for $f_{a,b}$. From Figure 3, we see that for some values of $a$ and $b$, the Hénon map $f$ is quite reminiscent of the horseshoe map $h$.

Since the map $f$ is polynomial in $x$ and $y$, we can also think of $x$ and $y$ as being complex-valued. In this case, $f : \mathbb{C}^2 \to \mathbb{C}^2$ is a holomorphic diffeomorphism of $\mathbb{C}^2$. This is also in some sense a change in the map $f$, but all of the dynamics of $f$ restricted to $\mathbb{R}^2$ are contained in the dynamics of the maps on $\mathbb{C}^2$, so we can still learn about the original map by studying it on this larger domain.

We next make a few observations about $f$. First, note that $f$ is the composition $f = f_3 \circ f_2 \circ f_1$ of the three maps

$$
\begin{aligned}
f_1(x,y) &= (x, by), \\
f_2(x,y) &= (-y, x), \\
f_3(x,y) &= (x + (-y^2 + a),\ y)
\end{aligned}
\tag{5.1}
$$

For $0 < b < 1$, the images of $B$ under the maps $f_1$ and $f_2 f_1$ are depicted in Figure 4, while $f$ is depicted in Figure 3 with some $a > 0$.

From the composition of these functions, we can easily see that $f$ has constant Jacobian determinant $\det(DF) = b$. Moreover, when $b = 0$, $f$ reduces to a quadratic polynomial on $\mathbb{C}$.

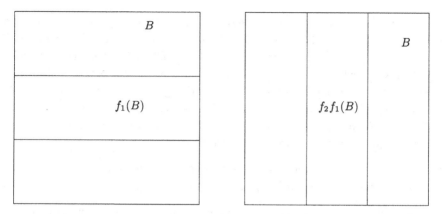

**Figure 4.** The Hénon map can be decomposed as $f = f_3 \circ f_2 \circ f_1$, where the component functions are defined in (5.1). Left: $f_1(B)$ sitting inside $B$. Right: $f_2 f_1(B)$ sitting inside $B$.

A simple argument shows that there is an $R = R(a, b)$ such that if we define the three sets

$$B = \{|x| < R, |y| < R\},$$
$$B^+ = \{|y| > R, |y| > |x|\},$$
$$B^- = \{|x| > R, |x| > |y|\},$$

then we have the same dynamical relations as for the corresponding sets for the horseshoe map. That is, points in $B^+$ can be mapped to $B^+$, $B$, or $B^-$, points in $B$ can be mapped to $B$ or $B^-$, and points in $B^-$ must be mapped to $B^-$.

Recall from Problem 2.1 that $K^+$ denotes the set of points with bounded forward orbit, $K^-$ the set of points with bounded backward orbit, and $K$ to be the intersection of these two sets.

When $a$ is large, the tip of $f_{a,b}(B)$ is outside $B$. For certain larger values of $a$, Devaney and Nitecki [1979] proved that $f_{a,b}$ "is" a horseshoe. By this we mean that $f$ is hyperbolic on the set $X$ of points that remain in $B$ for all time, and the dynamics of $f$ restricted to $X$ are topologically conjugate to those on the standard horseshoe of Section 4. Using complex techniques, Oberste-Vorth [1987] improved this result by showing that it works for any $a$ such that the tip of $f_{a,b}(B)$ is outside of $B$.

In Theorem 14.3 we will describe an optimal result in this direction.

EXAMPLE 5.1. To compare the dynamics of $f$ in the real and complex cases, consider $f_{a,b}$ with $a$ and $b$ real. As an ad hoc definition, let $K_{\mathbb{R}}$ be the set of $p \in \mathbb{R}^2$ with bounded forward and backward orbits, and let $K_{\mathbb{C}}$ be the set of $p \in \mathbb{C}^2$ with bounded forward and backward orbits. Then another result of Oberste-Vorth [1987] is that $K_{\mathbb{C}} = K_{\mathbb{R}}$.

Thus we already have a mental picture of $K$ for these parameter values. We can also get a picture of $K^+$ and $K^-$ in the complex case, since we can extend the analogy between $f$ and the horseshoe map by replacing the square $B$ by a bidisk $B = D(R) \times D(R)$ contained in $\mathbb{C}^2$, where $D(R)$ is the disk of radius $R$ centered at 0 in $\mathbb{C}$. In the definitions of $B^+$ and $B^-$, we can interpret $x$ and $y$ as complex-valued, in which case the definitions of these sets still make sense. Moreover, the same mapping relations hold among $B^+$, $B^-$, and $B$ as before. In this case, $B \cap K^+$ is topologically equivalent to the set product of a Cantor set and a disk, $B \cap K^-$ is equivalent to the product of a disk and a Cantor set, and $B \cap K$ is equivalent to the product of two Cantor sets.

THESIS. *A surprising number of properties of the horseshoe (when properly interpreted) hold for general complex Hénon diffeomorphisms.*

The "surprising" part of the above thesis is that the horseshoe map was designed to be simple and easily understood, yet it sheds much light on the less immediately accessible Hénon maps.

## 6. Properties of Horseshoe and Hénon Maps

We again consider some properties of the horseshoe map in terms of its periodic points. The investigation of periodic points plays an important role in the study of many dynamical systems. In Poincaré's words,

> What renders these periodic points so precious to us is that they are, so to speak, the only breach through which we might try to penetrate into a stronghold hitherto reputed unassailable.

As an initial observation, recall that from symbolic dynamics, we know that the periodic points are dense in $K$. In fact, it is not hard to show that these periodic points are all *saddle points;* that is, if $p$ has period $n$, then $(Dh^n)(p)$ has one eigenvalue larger than 1 in modulus, and one smaller. After recalling if necessary the definition of stable and unstable manifolds from (3.1) and (3.2), you should attempt to prove the following fact:

EXERCISE 6.1. For any periodic saddle point of the horseshoe map $h$, $W^s(p)$ is dense in $K^+$ and $W^u(p)$ is dense in $K^-$.

Now suppose $p \in K^+$, and let $n \in \mathbb{N}$ and $\varepsilon > 0$. By exercise 4.2, there is a periodic point $q$ with period $n$, and by this last exercise, the stable manifold for $q$ comes arbitrarily close to $p$. In particular, we can find $p' \in W^s(q)$ with $d(p, p') < \varepsilon$. Hence in any neighborhood of $p$, there are points that are asymptotic to a periodic point of any given period. We can contrast this with a point $p$ in the basin of attraction for a sink. In this case, for a small enough neighborhood of $p$, every point will be asymptotic to the same periodic point.

This example illustrates the striking difference between regular and chaotic behavior. In the case of a sink, the dynamics of the map are relatively insensitive to the precise initial conditions, at least within the basin of attraction.

But in the horseshoe case, the dynamics can change dramatically with an arbitrarily small change in the initial condition. In a sense, chaotic behavior occurs throughout $K^+$.

A second basic example of Axiom A behavior is the solenoid [Bowen 1978, p. 4]. Take a solid torus in $\mathbb{R}^3$ and map it inside itself so that it wraps around twice. The image of this new set then wraps around 4 times. The solenoid is the set that is the intersection of all the forward images of this map. Moreover, the map extends to a diffeomorphism of $\mathbb{R}^3$ and displays chaotic behavior on the solenoid, which is the attractor for the diffeomorphism.

EXAMPLE 6.2. Consider $f_{a,b} : \mathbb{C}^2 \to \mathbb{C}^2$ when $a$ and $b$ are small. It is shown in [Hubbard and Oberste-Vorth 1995] that $f_{a,b}$ has both a fixed sink and an invariant set having the topological and dynamical structure of a solenoid, so that it displays both regular and chaotic behavior in different regions.

Note that if $q$ is a sink, then $W^s(q) \subseteq K^+$ is open, and hence $W^s(q) \subseteq \mathring{K}^+$. On the interior of $K^+$, there is no chaos. To see this, suppose $p \in \mathring{K}^+$, and choose $\varepsilon > 0$ such that $\overline{\mathbb{B}_\varepsilon(p)} \subseteq K^+$. A simple argument using the form of $f$ and the definitions of $B$, $B^+$, and $B^-$ shows that any point in $K^+$ must eventually be mapped into $B$. Hence by compactness, there is an $n$ sufficiently large that $f^n(\overline{\mathbb{B}_\varepsilon(p)}) \subseteq B$. Since $B$ is bounded, we see by Cauchy's integral formula that the norm of the derivatives of $f^n$ are uniformly bounded on $\mathbb{B}_\varepsilon(p)$ independently of $n \geq 0$. This is incompatible with chaotic behavior. For more information and further references, see [Bedford and Smillie 1991b].

To start our study of sets where chaotic behavior can occur, we define $J^+ := \partial K^+$ and $J^- := \partial K^-$, where $K^+$ and $K^-$ are as in problem 2.1. The following theorem gives an analog of exercise 6.1 in the case of a general complex Hénon mapping, and is contained in [Bedford and Smillie 1991a].

THEOREM 6.3. *If $p$ is a periodic saddle point of the Hénon map $f$, then $W^s(p)$ is dense in $J^+$, and $W^u(p)$ is dense in $J^-$.*

We will see in Corollary 13.4 that a Hénon map $f$ has saddle periodic points of all but finitely many periods, so just as in the argument after exercise 6.1, we see that chaotic behavior occurs throughout $J^+$, and a similar argument applies to $J^-$ under backward iteration.

## 7. Dynamically Defined Measures

In the study of dynamics in one variable, there are many tools available coming from classical complex analysis, potential theory, and the theory of quasiconformal mappings. In higher dimensions, not all of these tools are available, but one tool that remains useful is potential theory. The next section will provide some background for the ways in which this theory can be used to study dynamics; but before talking about potential theory proper, we first discuss some measures associated with the horseshoe map $h$. With notation as in section 4, we define the level-$n$ set of $h$ to be the set $h^{-n}B \cap h^n B$. Since the forward images of $B$ are horizontal strips and the backward images of $B$ are vertical strips, we see that the level-$n$ set consists of $2^{2n}$ disjoint boxes.

ASSERTION. *For $j$ sufficiently large, the number of fixed points of $h^j$ in a component of the level-n set of h is independent of the component chosen.*

In fact, there is a unique probability measure $m$ on $K$ that assigns equal weight to each level-$n$ square, and the above assertion can be rephrased in terms of this measure. Let $P_k$ denote the set of $p \in \mathbb{C}^2$ such that $h^k(p) = p$. Then it follows from the above assertion that

$$\frac{1}{2^k} \sum_{p \in P_k} \delta_p \to m \tag{7.1}$$

in the topology of weak convergence.

We can use a similar technique to study the distribution of unstable manifolds. Again we consider the horseshoe map $h$, and we suppose that $p_0$ is a fixed saddle point of $h$ and that $S$ is the component of $W^u(p_0) \cap B$ containing $p_0$. In this case, $S$ is simply a horizontal line segment through $p_0$. Next, let $T$ be a line segment from the top to the bottom of $B$ so that $T$ is transverse to every horizontal line. See Figure 5. From the discussion of $h$, we know that $h^n(S) \cap B$ consists of $2^n$ horizontal line segments, so $h^n(S)$ intersects $T$ in $2^n$ points.

We can define a measure on $T$ using an averaging process as before. This time we average over points in $h^n(S) \cap T$ to obtain a measure $m_T^-$. Thus we have

$$\frac{1}{2^n} \sum_{p \in h^n(S) \cap T} \delta_p \to m_T^-, \tag{7.2}$$

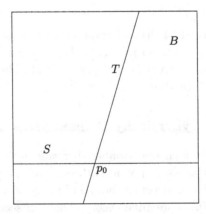

**Figure 5.** $S$ is a component of $W^u(p_0) \cap B$, and $T$ is a line transversal to all such components.

where again the convergence is in the weak sense. This gives a measure on $T$ which assigns equal weight to each level-$n$ segment; i.e., to each component of $h^n(B) \cap T$.

Note that if $T'$ is another segment like $T$, the unstable manifolds of $h$ give a way to transfer the above definition to a measure on $T'$. That is, given a point $p \in h^n(S) \cap T$, we can project along the component of $h^n(S) \cap B$ containing $p$ to obtain a point $p' \in T'$. Using this map we obtain a measure $\varphi(m_T^-)$ on $T'$. It is straightforward to show that this is the same measure as $m_{T'}^-$ obtained by using $T'$ in place of $T$ in (7.2).

This family of measures on transverse segments is called a *transversal measure* and we denote it by $m^-$. Using an analogous construction with stable manifolds, we can likewise define a measure $m^+$ defined on "horizontal" segments.

Finally, we can take the product of these two measures to get a measure $m = m^- \times m^+$ defined on $B$. Then one can show that this product measure is the same as the measure $m$ defined in (7.1). Hence there are at least two dynamically natural ways to obtain this measure.

The two transversal measures $m^+$ and $m^-$ and the product measure $m$ can be defined for general Axiom A diffeomorphisms [Ruelle and Sullivan 1975].

## 8. Potential Theory

To provide some physical motivation for the study of potential theory, consider two electrons moving in $\mathbb{R}^d$, each with a charge of $-1$. Then the repelling force between them is proportional to $1/r^{d-1}$. If we fix one electron at the origin, the total work in moving the other electron from the point $z_0$ to the point $z_1$ is independent of the path taken and is given by $P(z_1) - P(z_0)$, where $P$ is a potential function that depends on the dimension:

$$
\begin{cases}
P(z) = |z| & \text{if } d = 1, \\
P(z) = \log |z| & \text{if } d = 2, \\
P(z) = -\|z\|^{-(d-2)} & \text{if } d \geq 3.
\end{cases}
\tag{8.1}
$$

From the behavior of $P$ at 0 and $\infty$ we see that, if $d \leq 2$, the amount of work needed to bring a unit charge in from the point at infinity is infinite, while this work is finite for $d \geq 3$. On the other hand, if $d \geq 2$, the amount of work needed to bring two electrons together is infinite, but for $d = 1$ this work is finite.

We can think of a collection of electrons as a charge, and we can represent charges by measures $\mu$ on $\mathbb{R}^d$. Then, for $S \subseteq \mathbb{R}^d$, the amount of charge on $S$ is $\mu(S)$.

EXAMPLE 8.1. A unit charge at the point $z_0$ corresponds to the Dirac delta mass $\delta_{z_0}$.

By using measures to represent charges, we can use convolution to define potential functions for general charge distributions. That is, given a measure $\mu$ on $\mathbb{R}^d$, we define

$$P_\mu(z) = \int_{\mathbb{R}^d} P(z - w) \, d\mu(w), \qquad (8.2)$$

where $P$ is the appropriate potential function from (8.1). Note that this definition agrees with the previous definition of potential functions in the case of point charges. Note also that the assignment $\mu \mapsto P_\mu$ is linear in $\mu$.

In order to be able to use potential functions to study dynamics, we first need to understand a little more about their properties. In particular, we would like to know which functions can be the potential function of a finite measure.

In the case $d = 1$, the definition of $P_\mu$ and the triangle inequality imply that potential functions are convex, hence also continuous. We also have $P_\mu(x) = c|x| + O(1)$, where $c = \mu(\mathbb{R})$. In fact, any function $f$ satisfying these two conditions is a potential function of some measure. Hence a natural question is: How do we recover the measure from $f$?

In particular, given a convex function $f$ of one real variable, we can consider the assignment

$$f \mapsto \frac{1}{2} \left( \frac{\partial^2}{\partial x^2} f \right) dx, \qquad (8.3)$$

where the right-hand side is interpreted in the sense of distributions. By convexity, this distribution is positive. That is, it assigns a positive number to positive test functions, and a positive distribution is actually a positive measure. Hence we have an explicit correspondence between convex functions and positive measures, and with the additional restriction on the growth of potential functions given in the previous paragraph, we have an explicit correspondence between potential functions and finite positive measures. (The $\frac{1}{2}$ in the above formula occurs because we have normalized by dividing by the volume of the unit sphere in $\mathbb{R}$, which consists of the two points 1 and $-1$.)

In the case $d = 2$, the integral definition of $P_\mu$ implies that potential functions satisfy the *subaverage property*. That is, given a potential function $f$, any $z_0$ in the plane, and a disk $D$ centered at $z_0$, the value $f(z_0)$ is bounded above by the average of $f$ on $\partial D$. That is, if $\sigma$ represents one-dimensional Legesgue measure normalized so that the unit circle has measure 1, and if $r$ is the radius of $D$, then

$$f(z_0) \leq \frac{1}{r} \int_{\partial D} f(\zeta) \, d\sigma(\zeta).$$

Moreover, (8.2) implies that potential functions are *upper-semicontinuous;* a real-valued function $f$ is said to be upper-semicontinuous if its sub-level sets $f^{-1}(-\infty, a)$, for all $a \in \mathbb{R}$, are open. A function that is upper-semicontinuous and satisfies the subaverage property is called *subharmonic.*

Finally, if $f$ is subharmonic and satisfies $f(z) = c \log |z| + O(1)$ for some $c > 0$, then $f$ is said to be a *potential function.* Just as before, a potential function has the form $P_\mu$ for some measure $\mu$.

In fact, if $f$ is subharmonic and of class $C^2$, the Laplacian of $f$ is always posi-tive. This is an analog of the fact that the second derivative of a convex function is positive. If $f$ is subharmonic but not $C^2$, then $\Delta f$ is a positive distribution, hence a positive measure. Thus the Laplacian gives us a correspondence between potential functions and finite measures much like that in (8.3):

$$f \mapsto \frac{1}{2\pi}(\Delta f)\, dx\, dy,$$

where this is to be interpreted in the sense of distributions and again we have normalized by dividing by the volume of the unit sphere.

EXAMPLE 8.2. Applying the above assignment to the function $\log|z|$ produces the delta mass $\delta_0$ in the sense of distributions.

Now suppose that $K \subset \mathbb{R}^2 = \mathbb{C}$ is compact. Put a unit charge on $K$ and allow it to distribute itself through so that the mutual repulsion of the electrons is minimized. The distribution of charge on $K$ is described by a measure $\mu$. We will find $\mu$ by finding its potential function $P_\mu$, which is usually written $G$. The function $G$ satisfies the following properties:

1. $G$ is subharmonic.
2. $G$ is harmonic outside $K$.
3. $G = \log|z| + O(1)$.
4. $G$ is constant on $K$.

If $G$ satisfies properties 1 through 3, and also property

4$'$. $G \equiv 0$ on $K$,

we say that $G$ is a *Green function* for $K$. If $G$ exists, it is unique, and in this case we can take the Laplacian of $G$ in the sense of distributions. Thus, we say that

$$\mu_K := \frac{1}{2\pi}\Delta G\, dx\, dy$$

is the *equilibrium measure* for $K$.

EXAMPLE 8.3. Let $D$ be the unit disk. Then the Green function for $D$ is

$$G(z) = \log^+|z|,$$

where $\log^+|z| := \max\{\log|z|, 0\}$, and the equilibrium measure is

$$\mu_D = \frac{1}{2\pi}(\Delta \log^+|z|)\, dx\, dy,$$

which is simply arc length measure on $\partial D$, normalized to have mass 1.

## 9. Potential Theory in One-Variable Dynamics

In this section we discuss some of the ways in which potential theory can be used to understand the dynamics of polynomial maps of the complex plane. These ideas were introduced in [Brolin 1965]. In the next section we will see how they help us understand higher-dimensional complex dynamics.

For this section, let $f$ be a monic polynomial in one variable of degree $d \geq 2$, and let $K \subseteq \mathbb{C}$ be the set of $z$ such that the forward orbit of $z$ is bounded. Then $K$ has a Green function $G_K$, given by the formula

$$G_K(z) = \lim_{n \to \infty} \frac{1}{d^n} \log^+ |f^n(z)|.$$

It is difficult to understand Brolin's paper without knowing this formula. However, it was in fact first written down by Sibony in his UCLA lecture notes after Brolin's paper had already been written.

It is not hard to show that the limit in the definition of $G_K$ converges uniformly on compact sets, and since each of the functions on the right-hand side is subharmonic, the limit is also subharmonic. Moreover, on a given compact set outside of $K$, each of these functions is harmonic for sufficiently large $n$, so that the limit is harmonic on the complement of $K$. The property $G = \log |z| + O(1)$ follows by noting that for $|z|$ large we have $|z|^d/c \leq |f(z)| \leq c|z|^d$ for some $c > 1$, then taking logarithms and dividing by $d$, then using an inductive argument to bound $\left| \log^+ |f^n(z)|/d^n - \log |z| \right|$ independently of $d$. Finally, the property $G \equiv 0$ on $K$ is immediate since $\log^+ |z|$ is bounded for $z \in K$. In fact, $G_K$ has the additional property of being continuous.

Hence we see that $G_K$ really is the Green function for $K$, and we can define the equilibrium measure

$$\mu := \mu_K = \frac{1}{2\pi} (\Delta G_K) \, dx \, dy.$$

The following theorem provides a beautiful relationship between the measure $\mu$ and the dynamical properties of $f$. It says that we can recover $\mu$ by taking the average of the point masses at the periodic points of period $n$ and passing to the limit or by taking the average of the point masses at the inverse images of any nonexceptional point and passing to the limit. (A point $p$ is said to be *nonexceptional* for a polynomial $f$ if the set $\{f^{-n}(p) : n \geq 0\}$ contains at least three points. It is a theorem that there is at most one exceptional point for any polynomial.)

THEOREM 9.1 [Brolin 1965; Tortrat 1987]. *Let $f$ be a monic polynomial of degree $d$, and let $c \in \mathbb{C}$ be a nonexceptional point. Then*

$$\mu = \lim_{n \to \infty} \frac{1}{d^n} \sum_{z \in A_n} \delta_z,$$

*in the sense of convergence of measures, where $A_n$ is either the set of $z$ satisfying $f^n(z) = c$ (counted with multiplicity), or the set of $z$ satisfying $f^n(z) = z$ (counted with multiplicity).*

PROOF. We prove only the case $f^n(z) = c$ here. Let

$$\mu_n = \frac{1}{d^n} \sum_{f^n(z)=c} \delta_z.$$

Then we want to show that $\mu_n \to \mu_K$. Since the space of measures with the topology of weak convergence is compact, it suffices to show that if some subsequence of $\mu_n$ converges to a measure $\mu^*$, then $\mu^* = \mu_K$. By renaming, we may assume that $\mu_n$ converges to $\mu^*$.

We can show $\mu^* = \mu_K$ by showing the convergence of the corresponding potential functions. The potential function for $\mu_n$ is

$$G_n(z) = \frac{1}{d^n} \sum_{f^n(w)=c} \log |z - w| = \frac{1}{d^n} \log \left| \prod_{f^n(w)-c=0} (z - w) \right| = \frac{1}{d^n} \log |f^n(z) - c|.$$

Here the sum and products are taken over the indicated sets with multiplicities, and the last equality follows from the fact that we are simply multiplying all the monomials corresponding to roots of the monic polynomial $f^n(z) - c$.

Let $G^*(z) := \lim_{n\to\infty} G_n(z)$. Then $G^*$ is the potential function for $\mu^*$, and

$$G^*(z) = \lim_{n\to\infty} \frac{1}{d^n} \log |f^n(z) - c|,$$

while

$$G_K(z) = \lim_{n\to\infty} \frac{1}{d^n} \log^+ |f^n(z)|,$$

and we need to show that $G^*(z) = G_K(z)$. If $z \notin K$, then $f^n(z)$ tends to $\infty$ as $n$ increases, so that $G^*(z) = G_K(z)$ in this case. Since $G^*$ is the potential function for $\mu^*$, it is upper-semicontinuous, so it follows that $G^*(z) \geq 0$ for $z \in \partial K$. On the other hand, since $G^* = G$ on the set where $G = \varepsilon$, the maximum principle for subharmonic functions implies that $G \leq \varepsilon$ on the region enclosed by this set. Letting $\varepsilon$ tend to 0 shows that $G^* \leq 0$ on $K$.

Finally, using some knowledge of the possible types of components for the interior of $K$, one can show that, if $c$ is nonexceptional, the measure $\mu^*$ assigns no mass to the interior of $K$. This implies that $G^*$ is harmonic on $K$, since $\mu^*$ is the Laplacian of $G^*$. Hence both the maximum and minimum principles apply to $G^*$ on $K$, which implies that $G^* \equiv 0$ on $K$.

Thus $G^* \equiv G_K$ and hence $\mu^* \equiv \mu_K$ as desired. $\qquad \square$

REMARK. This theorem provides an algorithm for drawing a picture of the Julia set $J$ for a polynomial $f$. Start with a nonexceptional point $c$, and compute points on the backward orbits of $c$. These points will accumulate on the Julia set for $f$, and by discarding points in the first several backwards iterates of $c$, we can obtain a reasonably good picture of the Julia set. This algorithm has the disadvantage that these backwards orbits tend to accumulate most heavily on points in $J$ that are easily accessible from infinity. That is, it favors points at which a random walk starting at infinity is most likely to land and avoids points such as inward pointing cusps.

EXERCISE 9.2. Since $G$ is harmonic both in the complement of $K$ and in the interior of $K$, we see that supp $\mu \subseteq J$, where $J = \partial K$ is the Julia set. Show that supp $\mu = J$. Hint: Use the maximum principle.

Note that an immediate corollary of this exercise and Theorem 9.1 is that periodic points are dense in $J$.

## 10. Potential Theory and Dynamics in Two Variables

In Theorem 9.1, we took the average of point masses distributed over either the set $\{z : f^n(z) = c\}$, or the set $\{z : f^n(z) = z\}$. In the setting of polynomial diffeomorphisms of $\mathbb{C}^2$, there are two natural questions motivated by these results.

(i) What happens when we iterate one-dimensional submanifolds (forwards or backwards)?

(ii) Is the distribution of periodic points described by some measure $\mu$?

In $\mathbb{C}$, we can loosely describe the construction of the measure $\mu$ as first counting the number of points in the set $\{z : f^n(z) = c\}$ or $\{z : f^n(z) = z\}$, then using potential theory to describe the location of these points.

Before we consider such a procedure in the case of question (i) for $\mathbb{C}^2$, we first return to the horseshoe map and recall the measure $m^-$ defined in Section 7. Suppose that $B$ is defined as in that section, that $p$ is a fixed point for the horseshoe map $h$, that $S$ is the component of $W^u(p) \cap B$ containing $p$, and that $T$ is a line segment from the top to the bottom of $B$ as before. Then orient $T$ and $S$ so that these orientations induce the standard orientation on $\mathbb{R}^2$ at the point of intersection of $T$ and $S$.

Now apply $h$ to $S$. Then $h(S)$ and $T$ will intersect in two points, one of which is the original point of intersection, and one of which is new. See Figure 6. Because of the form of the horseshoe map, the intersection of $h(S)$ and $T$ at the new point will not induce the standard orientation on $\mathbb{R}^2$, but rather the opposite orientation. In general, we can apply $h^n$ to $S$, then assign $+1$ to each point of intersection that induces the standard orientation, and $-1$ to each point that induces the opposite orientation. Unfortunately, the sum of all such points of intersection for a given $n$ will always be 0, so this doesn't give us a way to count these points of intersection.

A second problem with real manifolds is that the number of intersections may change with small perturbations of the map. For instance, if the map is changed so that $h(S)$ is tangent to $T$ and has no other other intersections with $T$, then for small perturbations $g$ near $h$, $g(S)$ may intersect $T$ in zero, one, or two points.

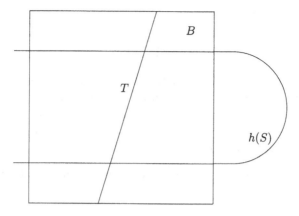

**Figure 6.** $T$ and $h(S)$.

Now suppose that $B$ is a bidisk in $\mathbb{C}^2$ (that is, the product of the disk with itself), that $h$ is a complex horseshoe map, and that $T$ and $S$ are complex submanifolds. In this case, there is a natural orientation on $T$ at any point given by taking a vector $v$ in the tangent space of $T$ over this point and using the set $\{v, iv\}$ to define the orientation at that point. We can use the same procedure on $S$, then apply $h^n$ as before. In this case, the orientation induced on $\mathbb{C}^2$ by $h^n(S)$ and $T$ is always the same as the standard orientation. Hence assigning $+1$ to such an intersection and taking the sum gives the number of points in $T \cap h^n(S)$.

Additionally, if both $S$ and $T$ are complex manifolds, the number of intersections between $h^n(S)$ and $T$, counted with multiplicity, is constant under small perturbations.

Thus, in studying question (i), we will use complex one-dimensional submanifolds.

Recall that, in the case of one variable, the Laplacian played a key role by allowing us to relate the potential function $G$ to the measure $\mu$. Here we consider an extension of the Laplacian to $\mathbb{C}^2$ in order to achieve a similar goal.

For a function $f$ of two real variables $x$ and $y$, the exterior derivative of $f$ is

$$df = \frac{\partial f}{\partial x} dx + \frac{\partial f}{\partial y} dy,$$

which is invariant under smooth maps. If we identify $\mathbb{R}^2$ with $\mathbb{C}$ in the usual way, multiplication by $i$ induces the map $(i)^*$ on the cotangent bundle, and this map takes $dx$ to $dy$ and $dy$ to $-dx$. Hence, defining $d^c = (i)^* d$, we have

$$d^c f = \frac{\partial f}{\partial x} dy - \frac{\partial f}{\partial y} dx,$$

which is invariant under smooth maps preserving the complex structure, that is, holomorphic maps. Hence $dd^c$ is also invariant under holomorphic maps.

Expanding $dd^c$ gives

$$dd^c f = d\left(\frac{\partial f}{\partial x}\right) dy - d\left(\frac{\partial f}{\partial y}\right) dx = \left(\frac{\partial^2 f}{\partial x^2} + \frac{\partial^2 f}{\partial y^2}\right) dx\, dy,$$

which is nothing but the Laplacian. This shows that the Laplacian, when viewed as a map from functions to two-forms, is invariant under holomorphic maps, and also shows that this procedure can be carried out in any complex manifold of any dimension. Moreover, it also shows that the property of being subharmonic is invariant under holomorphic maps.

EXERCISE 10.1. Let $D_r$ be the disk of radius $r$ centered at 0 in the plane, and compute

$$\int_{D_r} dd^c \log|z|.$$

(Hint: This is equal to $\int_{\partial D_r} d^c \log|z|$.)

We next need to extend the idea of subharmonic functions to $\mathbb{C}^2$.

DEFINITION 10.2. A function $f : \mathbb{C}^2 \to \mathbb{R}$ is *plurisubharmonic* if $h$ is upper-semicontinuous and if the restriction of $h$ to any one-dimensional complex line satisfies the subaverage property.

Intrinsically, an upper-semicontinuous function $h$ is plurisubharmonic if and only if $dd^c h$ is nonnegative, where again we interpret this in the sense of distributions.

In fact, in the above definition we could replace the phrase "complex line" by "complex submanifold" without changing the class of functions, since subharmonic functions are invariant under holomorphic maps. As an example of the usefulness of this and the invariance property of $dd^c$, suppose that $\varphi$ is a holomorphic embedding of $\mathbb{C}$ into $\mathbb{C}^2$ and that $h$ is smooth and plurisubharmonic on $\mathbb{C}^2$. Then we can either evaluate $dd^c h$ and pull back using $\varphi$, or we can first pull back and then apply $dd^c$. In both cases we get the same measure on $\mathbb{C}$.

## 11. Currents and Applications to Dynamics

In this section we provide a brief introduction to the theory of currents. A current is simply a linear functional on the space of smooth differential forms; that is, a current $\mu$ acts on a differential form of a given degree, say $\varphi = f_1\, dx + f_2\, dy$ in the case of a one-form, to give a complex number $\mu(\varphi)$, and this assignment is linear in $\varphi$. This is a generalization of a measure in the sense that a measure acts on zero-forms (functions) by integrating the function against the measure.

As an example, suppose that $M \subseteq \mathbb{C}^2$ is a submanifold of real dimension $n$. Then integration over $M$ is a current $[M]$ acting on $n$-forms $\varphi$; it is given simply by

$$[M](\varphi) = \int_M \varphi.$$

In this example the linearity is immediate, as is the relationship to measures. In particular, if $p \in \mathbb{C}^2$, then $[p] = \delta_p$, the delta mass at $p$, acts on zero-forms.

EXAMPLE 11.1. Suppose $P : \mathbb{C} \to \mathbb{C}$ is a polynomial having only simple roots, and let $R$ be the set of roots of $P$. Then $[R]$ is a current acting on 0-forms, and

$$[R] = \frac{1}{2\pi} dd^c \log |P|.$$

This formula is still true for arbitrary polynomials if we account for multiplicities in constructing $[R]$.

We can extend this last example to the case of polynomials from $\mathbb{C}^2$ to $\mathbb{C}$. This is the content of the next proposition.

PROPOSITION 11.2 (POINCARÉ–LELONG FORMULA). *If $P : \mathbb{C}^2 \to \mathbb{C}$ is a polynomial and $V = \{P = 0\}$, then*

$$[V] = \frac{1}{2\pi} dd^c \log |P|,$$

*where again $[V]$ is interpreted with weights according to multiplicity.*

Suppose now that $f : \mathbb{C}^2 \to \mathbb{C}^2$ is a Hénon diffeomorphism (see Section 5), and define

$$G^+(p) = \lim_{n \to \infty} \frac{1}{2^n} \log^+ |\pi_1(f^n(p))|,$$

$$G^-(p) = \lim_{n \to \infty} \frac{1}{2^n} \log^+ |\pi_2(f^{-n}(p))|,$$

where $\pi_j$ is projection to the $j$-th coordinate. As in the case of the function $G$ defined for a one-variable polynomial, it is not hard to check that $G^+$ is plurisubharmonic, is identically 0 on $K^+$, is pluriharmonic on $\mathbb{C}^2 \setminus K^+$ (i.e., is harmonic on any complex line), and is positive on $\mathbb{C}^2 \setminus K^+$. In analogy with the function $G$, we say that $G^+$ is the Green function of $K^+$. Likewise, $G^-$ is the Green function of $K^-$.

Note that for $n$ large and $p \notin K$, the value of $|\pi_1 f^n(p)|$ is comparable to the square of $|\pi_2 f^n(p)|$, and hence we may replace $|\pi_1 f^n(p)|$ by $\|f^n(p)\|$ in the formula for $G^+$, and likewise for $G^-$.

Again in analogy with the one-variable case, and using the equivalence between the Laplacian and $dd^c$ outlined earlier, we define

$$\mu^+ = \frac{1}{2\pi} dd^c G^+, \quad \mu^- = \frac{1}{2\pi} dd^c G^-.$$

Then $\mu^+$ and $\mu^-$ are currents supported on $J^+ = \partial K^+$ and $J^- = \partial K^-$, respectively. Moreover, $\mu^{\pm}$ restrict to measures on complex one-dimensional submanifolds in the sense that we can pull back $G^{\pm}$ from the submanifold to an open set in the plane, then take $dd^c$ on this open set.

As an analog of the case $f^n(z) = c$ of Theorem 9.1, we have the following theorem.

THEOREM 11.3. *Let $V$ be the (complex) $x$-axis in $\mathbb{C}^2$, i.e., the set where $\pi_2$ vanishes, and let $f$ be a complex Hénon map. Then*

$$\lim_{n \to \infty} \frac{1}{2^n} [f^{-n} V] = \mu^+.$$

PROOF. Note that the set $f^{-n} V$ is the set where the polynomial $\pi_1 f^n$ vanishes. Hence the previous proposition implies that

$$[f^{-n} V] = \frac{1}{2\pi} dd^c \log |\pi_1 f^n|.$$

Passing to the limit and using an argument like that in Theorem 9.1 to replace log by $\log^+$, we obtain the theorem. See [Bedford and Smillie 1991a] or [Fornæss and Sibony 1992a] for more details.                                      □

Here is a more comprehensive form of this theorem:

THEOREM 11.4. *If $S$ is a complex disk and $f$ is a complex Hénon map, then*

$$\lim_{n \to \infty} \frac{1}{2^n} [f^{-n} S] = c\mu^+,$$

*where $c = \mu^-[S]$. An analogous statement is true with $\mu^+$ and $\mu^-$ interchanged and $f^n$ in place of $f^{-n}$.*

As a corollary, we obtain the following theorem from [Bedford and Smillie 1991b].

COROLLARY 11.5. *If $p$ is a periodic saddle point, $W^u(p)$ is dense in $J^-$.*

PROOF. Replacing $f$ by $f^n$, we may assume that $p$ is fixed. Let $S$ be a small disk in $W^u(p)$ containing $p$. Then the forward iterates of $S$ fill out the entire unstable manifold. Moreover, by the previous theorem, the currents associated with these iterates converge to $c\mu^-$ where $c = \mu^+[S]$. If $c \neq 0$, the proof is complete since then $W^u(p)$ must be dense in $\operatorname{supp} \mu^- = J^-$.

But $c$ cannot be 0: if it were, $G^+|_S$ would be harmonic, hence identically 0 by the minimum principle since $G$ is nonnegative on $S$ and 0 at $p$. Hence $S$ would be contained in $K$, which is impossible since the iterates of $S$ fill out all of $W^u(p)$, which is not bounded. Thus $c \neq 0$, so $W^u(p)$ is dense in $J^-$.    □

This corollary gives some indication of why pictures of invariant sets on complex slices in $\mathbb{C}^2$ show essentially the full complexity of the map. If we start with any complex slice transverse to the stable manifold of a periodic point $p$, then the forward iterates of this slice accumulate on the unstable manifold of $p$, hence on all of $J^-$ by the corollary. All of this structure is then reflected in the original slice, giving rise to sets that are often self-similar and bear a striking resemblance to Julia sets in the plane.

## 12. Currents and Hénon Maps

In this section we continue the study of the currents $\mu^+$ and $\mu^-$ in order to obtain more dynamical information.

We first consider this in the context of the horseshoe map. Recall that $B$ is a square in the plane and that we have defined measures $m^+$ and $m^-$, and their product measure $m$, in Section 7.

In fact, $m^+$ and $m^-$ generalize to $\mu^+$ and $\mu^-$ in the case that the Hénon map is a horseshoe. More explicitly, let $D_\lambda$ be a family of complex disks in $\mathbb{C}^2$ indexed by the parameter $\lambda$, such that each $D_\lambda$ intersects $\mathbb{R}^2$ in a horizontal segment in $B$ and such that these segments fill out all of $B$. Then we can recover $\mu^+$, at least restricted to $B$, by

$$\mu^+|_B = \int [D_\lambda] \, dm^+(\lambda).$$

In analogy with the construction of $m$ as a product measure using $m^+$ and $m^-$, we would like to combine $\mu^+$ and $\mu^-$ to obtain a measure $\mu$. Since $\mu^+$ and $\mu^-$ are currents, the natural procedure to try is to take $\mu = \mu^+ \wedge \mu^-$. While forming the wedge product is not well-defined for arbitrary currents, it is well-defined in this case using the fact that these currents are obtained by taking $dd^c$ of a continuous plurisubharmonic function and applying a theorem of pluripotential theory. In this way we get a measure $\mu$ on $\mathbb{C}^2$.

DEFINITION 12.1. $\mu = \mu^+ \wedge \mu^-$.

DEFINITION 12.2. $J = J^+ \cap J^-$.

We next collect some useful facts about $\mu$.

(1) $\mu$ is a probability measure. For a proof of this, see [Bedford and Smillie 1991a].

(2) $\mu$ is invariant under $f$. To see this, note that, since

$$G^\pm = \lim_{n\to\infty} \frac{1}{2^n} \log^+ \|f^{\pm n}\|,$$

we have $G^\pm(f(p)) = 2^\pm G^\pm(p)$. Since $\mu^\pm = (1/2\pi)dd^c G^\pm$, this implies that $f^*(\mu^\pm) = 2^\pm \mu^\pm$, and hence

$$f^*(\mu) = f^*(\mu^+) \wedge f^*(\mu^-) = 2\mu^+ \wedge \tfrac{1}{2}\mu^- = \mu^+ \wedge \mu^- = \mu.$$

(3) $\operatorname{supp} \mu \subseteq J$. This is a simple consequence of the fact that the support of $\mu$ is contained in the intersection of $\operatorname{supp} \mu^+ = J^+$ and $\operatorname{supp} \mu^- = J^-$ and the definition of $J$.

In order to examine the support of $\mu$ more precisely, we turn our attention for a moment to Shilov boundaries. Let $X$ be a subset either of $\mathbb{C}$ or $\mathbb{C}^2$. We say that a set $B$ is a *boundary* for $X$ if $B$ is closed and if for any holomorphic polynomial $P$ we have

$$\max_X |P| = \max_B |P|.$$

With the right conditions, the intersections of any set of boundaries is a again a boundary by a theorem of Shilov, so we can intersect them all to obtain the smallest such boundary. This is called the *Shilov boundary* for $X$.

EXAMPLE 12.3. Let $X = D_1 \times D_1$, where $D_1$ is the unit disk. Then the Shilov boundary for $X$ is $(\partial D_1) \times (\partial D_1)$, while the topological boundary for $X$ is

$$\partial X = (D_1 \times \partial D_1) \cup (\partial D_1 \times D_1).$$

The following theorem is contained in [Bedford and Taylor 1987].

THEOREM 12.4. $\operatorname{supp} \mu = \partial_{\mathrm{Shilov}} K$.

We have already defined $J$ as the intersection of $J^+$ and $J^-$, and the choice of notation is designed to suggest an analogy with the Julia set in one variable. However, in two variables, the support of $\mu$ is also a natural candidate for a kind of Julia set. Hence we make the following definition.

DEFINITION 12.5. $J^* = \operatorname{supp} \mu$.

For subsets of $\mathbb{C}$ there is no distinction between the topological boundary and the Shilov boundary; hence we could have defined the Julia set $J$ as either the topological or the Shilov boundary of $K$.

## 13. Heteroclinic Points and Pesin Theory

In the previous section, we discussed some of the formal properties of $\mu$ arising from considerations of the definition and of potential theory. In this section we concentrate on the less formal properties of $\mu$ and on the relation of $\mu$ to periodic points. The philosophy here is that since $\mu^+$ and $\mu^-$ describe the distribution of one-dimensional objects, $\mu$ should describe the distribution of zero-dimensional objects.

An example of a question using this philosophy is the following. For a periodic point $p$, we know that $\mu^+$ describes the distribution of $W^s(p)$ and $\mu^-$ describes the distribution of $W^u(p)$. Does $\mu$ describe (in some sense) the distribution of intersections $W^s(p) \cap W^u(q)$?

DEFINITION 13.1. Let $p$ and $q$ be saddle periodic points of a diffeomorphism $f$. A point in the set $(W^u(p) \cap W^s(q)) \setminus \{p, q\}$ is called a *heteroclinic point*. If $p = q$, then such a point is called a *homoclinic point*.

Unfortunately, the techniques discussed so far do not allow us to prove the existence of even one heteroclinic point. We will imagine how it might be possible for the unstable manifold of $p$ to avoid the stable manifold of $q$. The stable and unstable manifolds are conformally equivalent to copies of $\mathbb{C}$, so we have parametrizations $\varphi_u : \mathbb{C} \to W^u(p)$ and $\varphi_s : \mathbb{C} \to W^s(q)$.

Now, if $\pi_j$ represents projection onto the $j$-th coordinate, then $\pi_1 \varphi_u : \mathbb{C} \to \mathbb{C}$ is an entire function, and as such can have an omitted value. As an example, $\pi_1 \varphi_u(z)$ could be equal to $e^z$ and hence would omit the value 0. It could happen that there is a second saddle point $q$ such that $W^s(q)$ is the $x$-axis, in which case $W^u(p) \cap W^s(q) = \varnothing$.

At first glance, it may seem that this contradicts some of our earlier results. It might seem that Theorem 11.4 should imply that $W^u(p)$ intersects transversals which cross $J^-$, but in fact, that statement is a statement about convergence of distributions. Each of the distributions must be evaluated against a test function, and the test function must be positive on an open set. Thus there is still room for $W^u(p)$ and $W^s(p)$ to be disjoint.

Hence, in order to understand more about heteroclinic points, we need a better understanding of the stable and unstable manifolds. One possible approach is to use what is known as Ahlfors' three-island theorem. This theorem concerns entire maps $\psi : \mathbb{C} \to \mathbb{C}$. Roughly, it says that if we have $n$ open regions in the plane and consider their inverse images under $\psi$, then some fixed proportion of them will have an inverse image that is compact and that maps injectively under $\psi$ onto the corresponding original region.

If we apply this theorem to the map $\pi_1 \varphi_u$ giving $W^u(p)$, we can divide the plane into increasingly more and smaller islands, and we can do this in such a way that at each stage we obtain more of $W^u(p)$ as the injective image of regions in the plane. The result is that we get a picture of $W^u(p)$ which is locally laminar.

Since $W^u(p)$ is dense in $J^-$, this gives us one possible approach to studying $\mu^-$, and we can use a similar procedure to study $\mu^+$. However, recall that our goal here is to describe heteroclinic points. Thus in order for this approach to apply, we need to be able to get the disks for $\mu^+$ to intersect the disks for $\mu^-$. Unfortunately, we don't get any kind of uniformity in the disks using this approach, so getting this intersection is difficult.

An alternate approach is to use the theory of nonuniform hyperbolicity (see [Young 1995] for more information). This is an extension of parts of the hyperbolic theory to a very general situation. This theory applies to the measure $\mu$, which is to say that we have expanding and contracting directions at $\mu$-almost every point, though the expansion and contraction need not be uniform and these directions need not depend continuously on the point. This is enough to produce stable and unstable manifolds through $\mu$-almost every point.

We can then identify the stable and unstable manifolds obtained using this theory with the disks obtained in the previous nonuniform laminar picture to guarantee that we get intersections between stable and unstable manifolds and hence heteroclinic points. Putting all of this together, we obtain the following theorem, contained in [Bedford, Lyubich, and Smillie 1993a].

THEOREM 13.2. *$J^*$ is the closure of the set of all periodic saddle points, and also the closure of the union of all $W^u(p) \cap W^s(q)$ over all periodic saddles $p$ and $q$.*

This theorem can be viewed as an analog of the theorem in one variable dynamics saying that the Julia set is the closure of the repelling periodic points. For this reason, the set $J^*$ is perhaps a better analogue of the Julia set in the two dimensional case than is $J$.

Recall that $J^* = \partial_{\text{shilov}} K \subseteq \partial K = J$. When $f$ is an Axiom A diffeomorphism, it is a theorem that $J^* = J$. However, it is an interesting open question whether this equality holds in general. If it were the case that $J \neq J^*$, then there would be a saddle periodic point $q$ and another point $p$ such that $p \in \overline{W^s(q)} \cap \overline{W^u(q)}$, but $p \notin \overline{W^s(q) \cap W^u(q)}$.

In fact, using the ideas of Pesin theory, one can get precise information about the number of periodic points of a given period and how their distribution relates to the measure $\mu$. This is contained in the following theorem and corollary from [Bedford, Lyubich, and Smillie 1993b].

THEOREM 13.3. *Let $f : \mathbb{C}^2 \to \mathbb{C}^2$ be a complex Hénon map, and let $P_n$ be either the set of fixed points of $f^n$ or the set of saddle points of minimal period $n$. Then*

$$\lim_{n \to \infty} \frac{1}{2^n} \sum_{p \in P_n} \delta_p = \mu.$$

For the following corollary, let $P_n$ be the set of saddle points of $f$ of minimal period $n$, and let $|P_n|$ denote the number of points contained in this set.

COROLLARY 13.4. *There are periodic saddle points of all but finitely many periods. More precisely, we have $\lim_{n \to \infty} |P_n|/2^n = 1$.*

Recall that the horseshoe map had periodic points of all periods, so while we haven't achieved that result for general Hénon maps, we have still obtained a good deal of information about periodic points and heteroclinic points.

## 14. Topological Entropy

Recall that the horseshoe map is topologically equivalent to the shift map on two symbols. One could also ask if it is topologically equivalent to the shift on four symbols. That is, if $h$ is the horseshoe map defined on the square $B$, then $h(B) \cap h^{-1}(B)$ consists of four components, and we can label these components with four symbols. However, with this labeling scheme, one can check by counting that not all sequences of symbols correspond to an orbit of a point in the way that sequences of two symbols did. In fact, the number of symbol sequences of length 2 corresponding to part of an orbit is 8, and the number of such sequences of length 3 is 16. Allowing longer sequences and letting

$S(n)$ denote the number of sequences of length $n$ which correspond to part of an orbit, we obtain the formula

$$\lim_{n\to\infty} \frac{1}{n} \log S(n) = \log 2.$$

The number $\log 2$ is the *topological entropy* of the horseshoe map, and can be defined as the maximum growth rate over all finite partitions. In general, the shift map on $N$ symbols has entropy $\log N$ and, since entropy is a topological invariant, we see that all of these different shift maps are topologically distinct.

In the case of a general Hénon map, we have the following theorem, contained in [Smillie 1990].

THEOREM 14.1. *The topological entropy of a complex Hénon map is* $\log 2$.

Topological entropy is a useful idea because it is connected to many different aspects of polynomial diffeomorphisms. It is a measure of area growth and of the growth rate of the number of periodic points, both of which are closely related to the degree of the map as a polynomial. Moreover, it is related to measure-theoretic entropy in the sense that, for any probability measure $\nu$, the measure-theoretic entropy $h_\nu(f)$ is bounded from above by the topological entropy $h_{\text{top}}(f)$. Moreover, $\mu$ is the *unique* measure for which $h_\mu(f) = h_{\text{top}}(f)$ [Bedford, Lyubich, and Smillie 1993a].

We can also consider topological entropy for real Hénon maps, that is, $f_{a,b} : \mathbb{R}^2 \to \mathbb{R}^2$ as in Section 5 with $a, b \in \mathbb{R}$. In contrast to the theorem above, in this case we have $0 \le h_{\text{top}}(f_\mathbb{R}) \le \log 2$, and all values are possible. However, one can show that not all values are possible for Axiom A diffeomorphisms, but only logarithms of algebraic numbers [Milnor 1988]. Moreover, we also have the following theorem [Bedford, Lyubich, and Smillie 1993a].

THEOREM 14.2. *For a Hénon map $f$ with real coefficients, the following are equivalent.*

1. $h_{\text{top}}(f_\mathbb{R}) = \log 2$.
2. $J^* \subseteq \mathbb{R}^2$.
3. $K \subseteq \mathbb{R}^2$.
4. *All periodic points are real.*

*Moreover, these conditions imply that $J = J^*$.*

PROOF. Condition 1 implies that $f_\mathbb{R}$ has a measure $\mu'$ of maximal entropy with $\operatorname{supp} \mu' \subseteq \mathbb{R}^2$. By uniqueness we have $\mu' = \mu^*$, so $\operatorname{supp} \mu^* \subseteq \mathbb{R}^2$, thus giving condition 2.

Condition 2 implies that $J^* = \partial_{\text{Shilov}} K$ is contained in $\mathbb{R}^2$, which implies that $K$ is contained in $\mathbb{R}^2$. This gives condition 3, and in fact, since polynomials in $\mathbb{R}^2$ are dense in the set of continuous functions of $\mathbb{R}^2$, this also implies that $\partial_{\text{Shilov}} K = K$, and hence $J^* = K$ and thus $J^* = J$ since $J^* \subseteq J \subseteq K$.

Condition 3 immediately implies condition 4.

Condition 4 together with theorem 13.2 implies that $J^* \subseteq \mathbb{R}^2$, which implies that $\operatorname{supp} \mu^* \subseteq \mathbb{R}^2$, which implies condition 1.                    $\square$

These conditions are true for the set of real Hénon maps that are horseshoes. We can identify such maps with their parameter values in $\mathbb{R}^2$, in which case the set of horseshoe maps is an open set in $\mathbb{R}^2$. Since topological entropy is continuous for $C^\infty$ diffeomorphisms, we see that maps on the boundary of this set also satisfy the above conditions.

Let's call a real Hénon map that satisfies the conditions of Theorem 14.2 a *maximal entropy map*. Real horseshoes are maximal entropy maps. As we will see shortly, there are maximal entropy maps that are not horseshoes. The following result shows that any maximal entropy map is either Axiom A or fails to be Axiom A in a very specific way. Recall that a homoclinic tangency is an intersection of $W^u(p)$ and $W^s(p)$ at some point $q \neq p$ for some saddle point $p$. This intersection is a homoclinic tangency if the stable and unstable manifolds are tangent at $q$, and this is a quadratic tangency if the manifolds have quadratic contact at $q$. A diffeomorphism with a homoclinic tangency violates one of the defining properties of hyperbolicity and hence is not Axiom A.

THEOREM 14.3. *If the conditions in Theorem 14.2 hold, then*

(i) *periodic points are dense in $K$,*
(ii) *every periodic point is a saddle with expansion constants bounded below,*
(iii) *either $f$ is Axiom A or $f$ has a quadratic homoclinic tangency.*

Theorem 14.3 gives a picture of how the property of being a horseshoe is lost as the parameters change. (Recall that horseshoes are by definition Axiom A.) Suppose we have a one-parameter family $f_t$ of real Hénon maps that starts out as a horseshoe then loses the Axiom A property. The set of parameters for which the map is a horseshoe is open because of structural stability, so there is some first parameter value $t_0$ at which the map is not Axiom A. What happens at this parameter value? The function $h_{\text{top}}(f_\mathbb{R})$ is continuous, so $f_{t_0}$ is a maximal entropy map but it is not Axiom A. According to the previous theorem there are pieces of stable and unstable manifolds that have a quadratic tangency. Let us assume that for $t$ past $t_0$ the pieces of stable and unstable manifolds pull through each other. This means that the intersection point of the stable and unstable manifolds is moving out of $\mathbb{R}^2$ and into $\mathbb{C}^2$. This causes a decrease in the topological entropy $h_{\text{top}}(f_\mathbb{R})$. Since the topological entropy is continuous as a function in parameter space and is an invariant of topological conjugacy, we pass through uncountably many topological conjugacy classes as we vary the parameter. This presents a striking contrast to the horseshoe example, in which small variations in the parameters produce topologically conjugate diffeomorphisms.

## 15. Suggestions for Further Reading

We have presented here one point of view on complex dynamics in several variables. For other viewpoints on polynomial diffeomorphisms the reader can consult [Hubbard and Oberste-Vorth 1994; 1995; Fornæss and Sibony 1992a; 1994]. Other related directions include the study of rational maps on complex projective spaces [Fornæss and Sibony 1992b; Hubbard and Papadopol 1994; Ueda 1986; 1991; 1994; 1992; $\geq$ 1997]. There is also interesting work on non-polynomial diffeomorphisms of $\mathbb{C}^2$ by Buzzard [1995; 1997; $\geq$ 1997].

## References

[Bedford 1991] E. Bedford, "Iteration of polynomial automorphisms of $\mathbb{C}^2$", pp. 847–858 in *Proceedings of the International Congress of Mathematicians* (Kyoto, 1990), Math. Soc. Japan, Tokyo, 1991.

[Bedford, Lyubich, and Smillie 1993a] E. Bedford, M. Lyubich, and J. Smillie, "Polynomial diffeomorphisms of $\mathbb{C}^2$, IV: The measure of maximal entropy and laminar currents", *Invent. Math.* **112** (1993), 77–125.

[Bedford, Lyubich, and Smillie 1993b] E. Bedford, M. Lyubich, and J. Smillie, "Distribution of periodic points of polynomial diffeomorphisms of $\mathbb{C}^2$", *Invent. Math.* **114** (1993), 277–288.

[Bedford and Smillie 1991a] E. Bedford and J. Smillie, "Polynomial diffeomorphisms of $\mathbb{C}^2$: currents, equilibrium measure and hyperbolicity", *Invent. Math.* **103** (1991), 69–99.

[Bedford and Smillie 1991b] E. Bedford and J. Smillie, "Polynomial diffeomorphisms of $\mathbb{C}^2$, II: Stable manifolds and recurrence", *J. Amer. Math. Soc.* **4** (1991), 657–679.

[Bedford and Smillie 1992] E. Bedford and J. Smillie, "Polynomial diffeomorphisms of $\mathbb{C}^2$, III: Ergodicity, exponents and entropy of the equilibrium measure", *Math. Ann.* **294**:3 (1992), 395–420.

[Bedford and Smillie 1997] E. Bedford and J. Smillie, "Quasihyperbolic polynomial diffeomorphisms of $\mathbb{C}^2$", 1997. Preprint, Cornell University.

[Bedford and Taylor 1987] E. Bedford and B. A. Taylor, "Fine topology, Šilov boundary, and $(dd^c)^n$", *J. Funct. Anal.* **72**:2 (1987), 225–251.

[Benedicks and Carleson 1991] M. Benedicks and L. Carleson, "The dynamics of the Hénon map", *Ann. of Math.* (2) **133** (1991), 73–169.

[Benedicks and Young 1993] M. Benedicks and L. S. Young, "Sinaĭ-Bowen-Ruelle measures for certain Hénon maps", *Invent. Math.* **112** (1993), 541–576.

[Bowen 1978] R. Bowen, *On Axiom A diffeomorphisms*, Regional Conference Series in Mathematics **35**, American Mathematical Society, Providence, R.I., 1978.

[Brolin 1965] H. Brolin, "Invariant sets under iteration of rational functions", *Ark. Mat.* **6** (1965), 103–144.

[Buzzard 1995] G. Buzzard, *Persistent homoclinic tangencies and infinitely many sinks for automorphisms of $\mathbb{C}^2$*, Ph.D. thesis, University of Michigan, 1995.

[Buzzard 1997] G. Buzzard, "Infinitely many periodic attractors for holomorphic automorphisms of two variables", *Annals of Math.* **145** (1997).

[Buzzard ≥ 1997]  G. Buzzard, "Kupka–Smale theorem for automorphisms of $\mathbb{C}^n$". To appear in *Duke J. Math.*

[Carleson and Gamelin 1993]  L. Carleson and T. W. Gamelin, *Complex dynamics*, Universitext: Tracts in Mathematics, Springer, New York, 1993.

[Devaney and Nitecki 1979]  R. Devaney and Z. Nitecki, "Shift automorphisms in the Hénon mapping", *Comm. Math. Phys.* **67** (1979), 137–146.

[Fornæss and Sibony 1992a]  J. E. Fornæss and N. Sibony, "Complex Hénon mappings in $\mathbb{C}^2$ and Fatou-Bieberbach domains", *Duke Math. J.* **65** (1992), 345–380.

[Fornæss and Sibony 1992b]  J. E. Fornæss and N. Sibony, "Critically finite rational maps on $\mathbb{P}^2$", pp. x+478 pp. in *The Madison Symposium on Complex Analysis* (Madison, WI, 1991), edited by A. Nagel and E. L. Stout, Contemporary Mathematics **137**, Amer. Math. Soc., Providence, 1992.

[Fornæss and Sibony 1994]  J. E. Fornæss and N. Sibony, "Complex dynamics in higher dimension. I", pp. 5, 201–231 in *Complex analytic methods in dynamical systems* (Rio de Janeiro, 1992), edited by C. Camacho et al., Astérisque **222**, Soc. math. France, Paris, 1994.

[Friedland and Milnor 1989]  S. Friedland and J. Milnor, "Dynamical properties of plane polynomial automorphisms", *Ergodic Theory Dynamical Systems* **9**:1 (1989), 67–99.

[Hubbard and Oberste-Vorth 1994]  J. H. Hubbard and R. W. Oberste-Vorth, "Hénon mappings in the complex domain, I: The global topology of dynamical space", *Inst. Hautes Études Sci. Publ. Math.* **79** (1994), 5–46.

[Hubbard and Oberste-Vorth 1995]  J. H. Hubbard and R. W. Oberste-Vorth, "Hénon mappings in the complex domain, II: Projective and inductive limits of polynomials", pp. 89–132 in *Real and complex dynamical systems* (Hillerød, 1993), edited by B. Branner and P. Hjorth, NATO Adv. Sci. Inst. Ser. C Math. Phys. Sci. **464**, Kluwer Acad. Publ., Dordrecht, 1995.

[Hubbard and Papadopol 1994]  J. H. Hubbard and P. Papadopol, "Superattractive fixed points in $\mathbb{C}^n$", *Indiana Univ. Math. J.* **43** (1994), 321–365.

[Milnor 1988]  J. Milnor, "Non-expansive Hénon maps", *Adv. in Math.* **69** (1988), 109–114.

[Oberste-Vorth 1987]  R. W. Oberste-Vorth, *Complex horseshoes and the dynamics of mappings of two complex variables*, Ph.D. thesis, Cornell University, 1987.

[Palis and Takens 1993]  J. Palis and F. Takens, *Hyperbolicity and sensitive chaotic dynamics at homoclinic bifurcations*, Cambridge Studies in Advanced Mathematics **35**, Cambridge University Press, Cambridge, 1993. Fractal dimensions and infinitely many attractors.

[Palis and de Melo 1982]  J. Palis, Jacob and W. de Melo, *Geometric theory of dynamical systems: An introduction*, Springer, New York, 1982.

[Ruelle 1989]  D. Ruelle, *Elements of differentiable dynamics and bifurcation theory*, Academic Press Inc., Boston, MA, 1989.

[Ruelle and Sullivan 1975]  D. Ruelle and D. Sullivan, "Currents, flows and diffeomorphisms", *Topology* **14** (1975), 319–328.

[Shub 1978]  M. Shub, *Stabilité globale des systèmes dynamiques*, Astérisque **56**, Soc. math. France, Paris, 1978. Translated as *Global stability of dynamical systems* by M. Shub, A. Fathi, and R. Langevin, Springer, New York, 1987.

[Smillie 1990]  J. Smillie, "The entropy of polynomial diffeomorphisms of $\mathbb{C}^2$", *Ergodic Theory Dynam. Systems* **10** (1990), 823–827.

[Suzuki 1992]  M. Suzuki, "A note on complex Hénon mappings", *Math. J. Toyama Univ.* **15** (1992), 109–121.

[Tortrat 1987]  P. Tortrat, "Aspects potentialistes de l'itération des polynômes", pp. 195–209 in *Séminaire de Théorie du Potentiel, Paris, No. 8*, edited by F. Hirsch and G. Mokobodzki, Lecture Notes in Math. **1235**, Springer, Berlin, 1987.

[Ueda 1986]  T. Ueda, "Local structure of analytic transformations of two complex variables, I", *J. Math. Kyoto Univ.* **26** (1986), 233–261.

[Ueda 1991]  T. Ueda, "Local structure of analytic transformations of two complex variables, II", *J. Math. Kyoto Univ.* **31** (1991), 695–711.

[Ueda 1992]  T. Ueda, "Complex dynamical systems on projective spaces", pp. 169–186 in *Topics around chaotic dynamical systems* (Kyoto, 1992), Sūrikaisekikenkyūsho Kōkyūroku **814**, Kyoto Univ., Kyoto, 1992. In Japanese.

[Ueda 1994]  T. Ueda, "Fatou sets in complex dynamics on projective spaces", *J. Math. Soc. Japan* **46** (1994), 545–555.

[Ueda ≥ 1997]  T. Ueda, "Critical orbits of holomorphic maps on projective spaces". To appear in *J. Geom. Anal.*

[Verjovsky and Wu 1996]  A. Verjovsky and H. Wu, "Hausdorff dimension of Julia sets of complex Hénon mappings", *Ergodic Theory Dynam. Systems* **16** (1996), 849–861.

[Wu 1993]  H. Wu, "Complex stable manifolds of holomorphic diffeomorphisms", *Indiana Univ. Math. J.* **42** (1993), 1349–1358.

[Yoccoz 1995]  J.-C. Yoccoz, "Introduction to hyperbolic dynamics", pp. 265–291 in *Real and complex dynamical systems* (Hillerød, 1993), edited by B. Branner and P. Hjorth, NATO Adv. Sci. Inst. Ser. C Math. Phys. Sci. **464**, Kluwer Acad. Publ., Dordrecht, 1995.

[Young 1995]  L.-S. Young, "Ergodic theory of differentiable dynamical systems", pp. 293–336 in *Real and complex dynamical systems (Hillerød, 1993)*, NATO Adv. Sci. Inst. Ser. C Math. Phys. Sci., 464, Kluwer Acad. Publ., Dordrecht, 1995.

# Index

JOHN SMILLIE
DEPARTMENT OF MATHEMATICS
CORNELL UNIVERSITY
ITHACA, NY 14853
  smillie@math.cornell.edu

GREGERY T. BUZZARD
DEPARTMENT OF MATHEMATICS
INDIANA UNIVERSITY
BLOOMINGTON, IN 47405
  gbuzzard@math.indiana.edu

Flavors of Geometry
MSRI Publications
Volume **31**, 1997

# Volume Estimates and Rapid Mixing

## BÉLA BOLLOBÁS

### Contents

## Lecture 1. Introduction

Computing, or at least estimating, the volume of a body is one of the oldest questions in mathematics, studied already in Egypt and continued by Euclid and Archimedes. Here we are mostly concerned with the *computational complexity* of estimating volumes of convex bodies in $\mathbb{R}^n$, with $n$ large. The problems and results to be discussed are all rather recent, most of them less than ten years old, and although there was a breakthrough a few years ago and by now there are several substantial and exciting results, there is much to be done.

In vague terms, we would like to find a fast algorithm that computes, for each convex body $K$ in $\mathbb{R}^n$, positive numbers $\underline{\text{vol}}\,K$ and $\overline{\text{vol}}\,K$ such that

$$\underline{\text{vol}}\,K \leq \text{vol}\,K \leq \overline{\text{vol}}\,K,$$

and $\overline{\text{vol}}\,K/\underline{\text{vol}}\,K$ is as small as possible.

This formulation has several flaws. It is not clear what our algorithm is allowed to do and how its speed is measured; we also have to decide how our convex body is given and how small $\overline{\text{vol}}\,K/\underline{\text{vol}}\,K$ we wish to make. Our first aim is then to make this problem precise.

We write $|x|$ for the standard Euclidean norm of a vector $x \in \mathbb{R}^n$, and $\langle x, y \rangle$ for the standard inner product. $B^n = B^n_2$ denotes the Euclidean ball of radius 1 in $\mathbb{R}^n$, and $B^n(\varepsilon)$ the ball of radius $\varepsilon$. If there is no danger of confusion, we write vol $K$ for the volume of a body $K$, in whatever the appropriate dimension is; if we want to draw attention to the dimension we write $\mathrm{vol}_k L$ for the $k$-dimensional volume of a $k$-dimensional body $L$.

By a *convex body* in $\mathbb{R}^n$ we mean a compact convex subset of $\mathbb{R}^n$, with nonempty interior. Following [Grötschel, Lovász, and Schrijver 1988], we shall assume that our convex body $K \subset \mathbb{R}^n$ is given by a certain oracle; this will enable us to obtain results that are valid for a large class of specific algorithms. An *oracle* is a "black box" that answers various questions put to it. When we talk about a convex body $K$ given by an oracle, the questions tend to be simple, for example: "What about the point $x \in \mathbb{R}^n$?" A *strong membership oracle* answers this question in one of two ways: "$x \in K$" or "$x \notin K$." A *strong separation oracle* answers either "$x \in K$" or, in addition to replying that "$x \notin K$," it gives a linear functional separating $x$ from $K$. In other words, when the answer is negative, the oracle justifies its assertion by displaying a vector $c \in \mathbb{R}^n$ such that $\langle c, x \rangle > \langle c, y \rangle$ for every $y \in K$. We can assume that $\|c\|_\infty = 1$, where $\|\cdot\|_\infty$ denotes the sup norm.

In a *weak membership oracle* the question is slightly different: "What about the point $x \in \mathbb{R}^n$ and the positive number $\varepsilon$?" Let

$$K_\varepsilon = K + B^n(\varepsilon) = \{y \in \mathbb{R}^n : |y - z| \leq \varepsilon \text{ for some } z \in K\}$$

and

$$K_{-\varepsilon} = \mathbb{R}^n \setminus (\mathbb{R}^n \setminus K)_\varepsilon = \{y \in \mathbb{R}^n : |y - z| > \varepsilon \text{ for every } z \notin K\};$$

the weak membership oracle answers either "$x \in K_\varepsilon$" or "$x \notin K_{-\varepsilon}$". (For $x \in K_\varepsilon \setminus K_{-\varepsilon}$ it may return either answer.)

Similarly, a *weak separation oracle* replies to the same question in one of the following two ways: either

$$\text{"}x \in K_\varepsilon\text{"}$$

or "$x \notin K_{-\varepsilon}$ and here is a functional $c \in \mathbb{R}^n$, $\|c\|_\infty = 1$, proving it:

$$\langle c, y \rangle < \langle c, x \rangle + \varepsilon \quad \text{for all } y \in K_{-\varepsilon}.\text{"}$$

An algorithm then is a sequence of questions to the oracle, each question depending on the answers to the previous questions. The *complexity* or *running time* of an algorithm is the number of questions asked before the bounds are produced.

A moment's thought tells us that with the oracles just described no algorithm can produce bounds other than $\underline{\mathrm{vol}}\,K = 0$ and $\overline{\mathrm{vol}}\,K = \infty$, since we can do no better than this if we imagine that $K$ is "at infinity" in the direction of the $x_1$-axis, say; if to the question: "What about $x = (x_i)_1^n$?" the oracle replies that

$$K \subset \{y = (y_i)_1^n : y_1 > x_1 + 1\},$$

then after some questions all we know is that $\min\{y_1 : (y_i) \in K\}$ is large, but we know nothing about the volume of the body. Indeed, we cannot even find a single point of $K$ in a finite time, even if $K$ is known to have large volume. In order to give the algorithm a chance, we need some guarantees about $K$, namely that it is not "at infinity" and it is not too small. The standard way to do this is to assume that

$$rB^n \subset K \subset RB^n \qquad (1.1)$$

for some positive numbers $r$ and $R$. If (1.1) holds we say that the algorithm is *well guaranteed*, with guarantees $r$ and $R$.

We should also specify how the data are measured: the *size* of the input $(r, R, n)$ for a convex body $K \subset \mathbb{R}^n$ satisfying (1.1) is

$$\langle K \rangle = n + \langle r \rangle + \langle R \rangle,$$

where $\langle x \rangle$ is the number of binary digits of a dyadic rational $x$.

In what follows, it will make very little difference which of the above oracles we shall use: the difficulty is not in going from one oracle to another but in finding suitable algorithms. For example, Grötschel, Lovász, and Schrijver [1988] showed that a weak separation oracle can be obtained from a weak membership oracle in polynomial time.

We are only interested in algorithms that are fairly fast, namely those whose complexity is polynomial and of not too high a degree. In view of this, our problem could be restated as follows. Given a polynomial $f(x) = x^a + b$, find a function $g(x)$ such that, if the oracle describing our convex body $K \subset \mathbb{R}^n$ has guarantees $2^{-l_1}$ and $2^{l_2}$, we can compute, after no more than $f(n + l_1 + l_2)$ appeals to the oracle, numbers $\underline{\mathrm{vol}}\, K$ and $\overline{\mathrm{vol}}\, K$ such that

$$\underline{\mathrm{vol}}\, K \le \mathrm{vol}\, K \le \overline{\mathrm{vol}}\, K$$

and $\overline{\mathrm{vol}}\, K / \underline{\mathrm{vol}}\, K \le g(n)$. Moreover, $g(x)$ should grow as slowly as possible.

As we shall see in the next section, the solution to this problem is rather disappointing: no matter what polynomial $f$ we choose, the approximation $g(n)$ cannot be guaranteed to be better than polynomial. However, if we do not insist that our approximations $\underline{\mathrm{vol}}\, K$ and $\overline{\mathrm{vol}}\, K$ be valid every time, we can do much better. In 1989, Dyer, Frieze, and Kannan [1991] devised a randomized algorithm that approximates with high probability the volume of a convex body as closely as desired in polynomial time. After this breakthrough, faster and faster algorithms have been devised, but it is unlikely that we are near to a best possible algorithm. In Section 4 we describe one of the most elegant (although not quite the fastest) algorithms found so far. This algorithm and every other are based on rapidly mixing random walks: we shall present the relevant results in Section 3.

## Lecture 2. Volumes of Convex Hulls
## and Deterministic Bounds

Our aim in this section is to show that there is no fast deterministic algorithm for approximating the volume of a convex body:

THEOREM 2.1. *For every polynomial-time algorithm for computing the volume of a convex body in $\mathbb{R}^n$ given by a well-guaranteed separation oracle, there is a constant $c > 0$ such that*

$$\frac{\overline{\mathrm{vol}\,K}}{\underline{\mathrm{vol}\,K}} \leq \left(\frac{cn}{\log n}\right)^n$$

*cannot be guaranteed for $n \geq 2$.*

Nevertheless, we start with a positive result, claiming that in polynomial time we *can* achieve some kind of approximation. To be precise, Lovász [1986] proved that for every convex body $K$ given by a well-guaranteed oracle there is an affine transformation $\varphi : x \to Ax + b$ computable in polynomial time and such that

$$B \subset \varphi(K) \subset n\sqrt{n+1}\,B.$$

In particular, this algorithm produces estimates $\underline{\mathrm{vol}}\,K$ and $\overline{\mathrm{vol}}\,K$ with

$$\overline{\mathrm{vol}}\,K/\underline{\mathrm{vol}}\,K \leq n^n(n+1)^{n/2}.$$

Furthermore, Lovász showed that if $K$ is centrally symmetric—say, if it is the unit ball of a norm on $\mathbb{R}^n$—there is a polynomial-time algorithm that produces estimates $\underline{\mathrm{vol}}\,K$ and $\overline{\mathrm{vol}}\,K$ with $\overline{\mathrm{vol}}\,K/\underline{\mathrm{vol}}\,K \leq n^n$.

Elekes [1986] was the first to realize that this bound is not as outrageous as it looks at first sight. Indeed, he showed that, for $0 < \varepsilon < 2$, there is no polynomial-time algorithm that returns $\overline{\mathrm{vol}}\,K$ and $\underline{\mathrm{vol}}\,K$ with

$$\overline{\mathrm{vol}}\,K/\underline{\mathrm{vol}}\,K \leq (2-\varepsilon)^n.$$

What Elekes noticed was that the convex hull of polynomially many points in $B^n$ is only a small fraction of $B^n$, and that this fact implies that every polynomial algorithm is bound to give a poor result. The theorem of Elekes was soon improved by Bárány and Füredi [1988] to an essentially best possible result: this is the main result we shall present.

Let's start with the problem of approximating the unit ball $B^n \subset \mathbb{R}^n$ by the convex hull of $m$ points of $B^n$. As we shall see later, a similar (and essentially equivalent) problem is that of approximating $B^n$ by the intersection of $m$ slabs, each containing $B^n$. A question of this type was considered in Section 2 of Keith Ball's lectures in this volume [Ball 1997], where the measure of approximation was the Banach–Mazur distance. Here our aim is rather different: we wish to approximate $B^n$ by polytopes contained by $B^n$ (or containing $B^n$) that have relatively few vertices, and the measure of our approximation is the difference of volumes.

The following beautiful and simple result of Elekes [1986] shows that we cannot hope for a good approximation unless we take exponentially many points. This result is reminiscent of [Ball 1997, Theorem 2.1], but the proof is considerably simpler.

THEOREM 2.2. *Let* $v_1, \ldots, v_m \in \mathbb{R}^n$ *and* $K = \mathrm{conv}\{v_1, \ldots, v_m\}$. *Then* $K \subset \bigcup_{i=1}^m B_i$, *where* $B_i = B(v_i/2, |v_i|/2)$ *is the ball of centre* $v_i/2$ *and radius* $|v_i|/2$. *In particular, if each* $v_i$ *is in the unit ball* $B^n$ *of* $\mathbb{R}^n$ *then*

$$\mathrm{vol}_n K / \mathrm{vol}_n B^n \le m/2^n.$$

PROOF. Suppose $x \notin B_i$, so that $\langle x - v_i/2, x - v_i/2 \rangle > \|v_i\|^2/4$. Then

$$|x|^2 > \langle x, v_i \rangle,$$

that is, $v_i$ is in the open half-space $H(x) = \{y \in \mathbb{R}^n : \langle x, y \rangle < \|x\|^2\}$. Hence if $x \notin \bigcup_{i=1}^m B_i$ then $v_i \in H(x)$ for every $i$, and so $K = \mathrm{conv}\{v_1, \ldots, v_m\} \subset H(x)$. But $x \notin H(x)$, so $x \notin K$. Hence $K \subset \bigcup_{i=1}^m B_i$, as claimed. The last statement of the theorem follows immediately. $\qquad \square$

To give a more geometric argument for why $K \subset \bigcup_{i=1}^m B_i$, note that $B_i$ is the set of points $x$ for which the angle $v_i \, x \, v_0$ is at least $\pi/2$, where $v_0 = 0 \in \mathbb{R}^n$. All we have to notice is that if $v = \lambda v_i + (1-\lambda)v_j$ for some $\lambda$ with $0 < \lambda < 1$, then $x \notin B_i \cup B_j$ implies that the angle $v \, x \, v_0$ is less than $\pi/2$, as the angles $v_i \, x \, v_0$ and $v_j \, x \, v_0$ are less than $\pi/2$. But then if $x \notin \bigcup_{i=1}^m B_i$ and $v \in K$ then the angle $v \, x \, v_0$ is less than $\pi/2$. Hence $\bigcup_{i=1}^m B_i$ contains the set

$$\alpha(K) = \{x \in \mathbb{R}^n : \text{angle } v \, x \, v_0 \ge \pi/2 \text{ for some } v \in K\},$$

which certainly contains $K$.

If $m$ is not too large the inequality of Theorem 2.2 is fairly good, but if $m$ is exponential, let alone at least $2^n$, it is very weak. Our next aim is to present a result of Bárány and Füredi [1988] (Theorem 2.5) that gives an essentially best possible bound.

Let's write $V(n, m)$ for the maximal volume of the convex hull of $m$ points in $B^n$:

$$V(n, m) = \max \{\mathrm{vol}_n K : K = \mathrm{conv}\{v_1, \ldots, v_m\} \subset B^n\},$$

and let

$$W(n, m) = \frac{V(n, m)}{\mathrm{vol}_n B^n}$$

be the proportion of the volume. In order to get our upper bound for $V(n, m)$, we need an extension of a result from [Fejes Tóth 1964] closely related to Fritz John's theorem [1948]. Theorem 3.1 in [Ball 1997] is a sharper version of Fritz John's theorem: here we shall need only the original result.

THEOREM 2.3. *For every convex body $K \subset \mathbb{R}^n$, there is a unique ellipsoid $E$ of maximal volume contained in $K$. If $E$ has centre $0$ then*

$$E \subset K \subset nE. \tag{2.1}$$

PROOF. By a simple compactness argument, there is at least one ellipsoid of maximal volume. To prove uniqueness, suppose that there are two ellipsoids of maximal volume, say $E$ and $E'$. Then the ellipsoid $\frac{1}{2}(E + E')$ is contained in the convex hull of $E \cup E'$. By the Brunn–Minkowski Theorem (or by the AM/GM inequality), $\mathrm{vol}_n\left(\frac{1}{2}(E + E')\right)$ is greater than $\mathrm{vol}_n E = \mathrm{vol}_n E'$, unless $E'$ is a translate of $E$. If $E'$ is a translate of $E$ we can assume without loss of generality that each is a unit ball, and then $\mathrm{conv}(E \cup E')$ is easily seen to contain an ellipsoid $E''$ with $\mathrm{vol}_n E'' > \mathrm{vol}_n E$. Hence $E$ is indeed unique.

To see that $K \subset nE$, all we have to check is that if $B^n$ is the unit ball in $\mathbb{R}^n$ and $|v| > n$ then $\mathrm{conv}(B^n \cup \{v\})$ contains another ellipsoid of volume $\mathrm{vol}_n B^n$. This can be done by simple calculations.                                       □

Let $S_0$ be a regular simplex with inscribed ball $B_0 = B^n$ and so with circumscribed ball $nB_0$. By the uniqueness of the ellipsoid of maximal volume in a convex body, $B_0$ is *the* ellipsoid of maximal volume in $S_0$. Also, it is simple to see that $S_0$ is a simplex of maximal volume contained in $nB_0$. Since the volume ratio is affine invariant, it follows that if $S$ is any simplex in $\mathbb{R}^n$ and $E_1 \subset S \subset E_2$ for some ellipsoids $E_1$ and $E_2$, then $\mathrm{vol}\, E_1/\mathrm{vol}\, E_2 \leq n^{-n}$. In particular, if $E \subset S \subset \lambda E$ for some ellipsoid $E$ and positive real $\lambda$ then $\lambda \geq n$. *A fortiori*, if $K$ is a simplex, we cannot replace $n$ in Theorem 2.3 by a smaller constant.

Also, if a simplex $S \subset \mathbb{R}^n$ contains a ball of radius $r_1$ and is contained in a ball of radius $r_2$, then $r_2 \geq nr_1$. This last assertion is the result from [Fejes Tóth 1964] that we shall need and extend.

In order to state and prove this extension, we introduce some notation. First, given a set $S \subset \mathbb{R}^n$, let $U = \mathrm{span}\, S = \mathrm{lin}\{x - y : x, y \in S\}$ be the subspace of $\mathbb{R}^n$ defined by $S$, and for $\rho > 0$ set

$$S^\rho = S + \left(U^\perp \cap \rho B^n\right),$$

where $U^\perp$ is the orthogonal complement of the subspace $U$. Thus $S^\rho$ is the set of $x \in \mathbb{R}^n$ for which $n(x, S)$ is attained at some $y \in S$ with $n(x, y) \leq \rho$ and $(x - y) \perp U$. If $S$ is convex and $\dim U = k$, then clearly

$$\mathrm{vol}_n S^\rho = \left(\mathrm{vol}_k S\right)\left(\mathrm{vol}_{n-k} B^{n-k}\right)\rho^{n-k}. \tag{2.2}$$

Secondly, for $1 \leq k \leq n$ define

$$\rho(n, k) = \begin{cases} 1 & \text{if } k = 0, \\ \sqrt{(n-k)/(nk)} & \text{if } 1 \leq k \leq n-2, \\ 1/n & \text{if } k = n-1. \end{cases}$$

LEMMA 2.4. *Let* $x \in S = \text{conv}\{v_0, v_1, \ldots, v_n\} \subset B^n$. *Then, for every* $k$ *such that* $0 \le k \le n-1$, *the simplex* $S$ *has a* $k$-*dimensional face* $S_k = \text{conv}\{v_{i_0}, v_{i_1}, \ldots, v_{i_k}\}$ *and a point* $x_k$ *in the interior of the face* $S_k$ *(in the sense that* $x_k = \sum_{j=0}^{k} \lambda_j v_{i_j}$ *with* $\sum_{j=0}^{k} \lambda_j = 1$ *and* $\lambda_j > 0$ *for every* $j$), *such that* $(x - x_k) \perp \text{span } S_k$ *and* $\|x - x_k\| \le \rho(n, k)$. *In particular,* $x \in S_k^{\rho(n, k)}$.

PROOF. We know the result for $k = n - 1$, and from it we shall deduce the result for $1 \le k \le n - 2$. Set $x_n = x$, $S_n = S$, and let $S_{n-1}$ be a $(n-1)$-dimensional face of $S_n$ containing a point $x_{n-1}$ such that $(x_n - x_{n-1}) \perp \text{span } S_{n-1}$ and $|x_n - x_{n-1}| \le 1/n$. Next, let $S_{n-2}$ be a $(n-2)$-dimensional face of $S_{n-1}$ containing a point $x_{n-2}$ such that $(x_{n-1} - x_{n-2}) \perp \text{span } S_{n-2}$ and $|x_{n-1} - x_{n-2}| \le 1/(n-1)$. Proceed in this way up to $x_k$ in $S_k$. Then the vectors $x_n - x_{n-1}$, $x_{n-1} - x_{n-2}$, ..., $x_{k+1} - x_k$ are orthogonal, with $|x_l - x_{l-1}| \le 1/l$. Hence

$$x - x_k = x_n - x_k = (x_n - x_{n-1}) + (x_{n-1} - x_{n-2}) + \cdots + (x_{k+1} - x_k)$$

is orthogonal to $\text{span } S_k$ and

$$|x - x_k|^2 \le \frac{1}{n^2} + \frac{1}{(n-1)^2} + \cdots + \frac{1}{(k+1)^2}$$

$$\le \frac{1}{n(n-1)} + \frac{1}{(n-1)(n-2)} + \cdots + \frac{1}{k(k+1)}$$

$$= \frac{1}{k} - \frac{1}{k+1} + \frac{1}{k+1} - \frac{1}{k+2} + \cdots + \frac{1}{n-1} - \frac{1}{n} = \frac{1}{k} - \frac{1}{n} = \frac{n-k}{nk},$$

as required.

Finally, let's consider the case $k = 0$. Suppose that $\|x - v_i\| > 1$ for every $i$, and so

$$\{v_0, v_1, \ldots, v_n\} \subset B^n \setminus B(x, 1).$$

But then every point of $S = \text{conv}\{v_0, v_1, \ldots, v_n\}$ is closer to 0 than to $x$ and so $x \notin S$. Hence $S \subset \bigcup_{i=0}^{n} B(v_i, 1)$, as claimed. $\qquad \square$

The alert reader must have noticed that this simple proof does not give a tight bound except in the cases $k = n$ and $k = 1$. This is because the later simplices $S_{n-1}, S_{n-2}, \ldots, S_{k+1}$ are likely to be contained in balls of radii less than 1. However, the loss is surprisingly little. Trivially, we cannot do better than in the case of a regular simplex inscribed in $B^n$: to cover the origin by neighbourhoods of the $k$-dimensional faces we must have $\rho(n, k)$ at least $\sqrt{(n-k)/(n(k+1))}$ rather than $\sqrt{(n-k)/(nk)}$. It is very likely that, in fact, $\rho(n, k) = \sqrt{(n-k)/(n(k+1))}$.

The Bárány–Füredi upper bound for $V(n, m)$ follows easily from Lemma 2.4:

THEOREM 2.5. *There is a constant* $c > 0$ *such that for* $m = m(n) \ge 1$ *we have*

$$W(n, m) \le \left( \frac{c(\log(m/n) + 1)}{n} \right)^{n/2}$$

*and so*

$$V(n,m) \leq \left( \frac{\gamma(\log(m/n) + 1)^{1/2}}{n} \right)^n,$$

*where $\gamma = (2\pi ec)^{1/2}$. Furthermore, if $\varepsilon > 0$ is fixed and we take $m/n \to \infty$ and $n/\log(m/n) \to \infty$, then*

$$V(n,m) \leq \left( \frac{(2e + \varepsilon)\log(m/n)}{n^2} \right)^{n/2}.$$

PROOF. Let $K = \mathrm{conv}\{v_1, v_2, \ldots, v_m\} \subset B^n$. By Carathéodory's theorem ([Carathéodory 1907]; see also [Eckhoff 1993]) $K$ is the union of its $n$-dimensional simplices:

$$K = \bigcup_{i_0 < \cdots < i_n} \mathrm{conv}\{v_{i_0}, v_{i_1}, \ldots, v_{i_n}\}.$$

Hence, by Lemma 2.4, for all $k$ such that $1 \leq k \leq n - 1$ we have

$$K \subset \bigcup_{i_0 < \cdots < i_k} \{S^{\rho(n,k)} : S = \mathrm{conv}\{v_{i_0}, \ldots, v_{i_k}\}\},$$

and so

$$\mathrm{vol}_n K \leq \binom{m}{k+1} \max\{\mathrm{vol}_n S^{\rho(n,k)} : S = \mathrm{conv}\{x_0, x_1, \ldots, x_k\} \subset B^n\}.$$

By identity (2.2), for a simplex $S$ as above,

$$\mathrm{vol}_n S^{\rho(n,k)} = (\mathrm{vol}_k S)(\mathrm{vol}_{n-k} B^{n-k}) \rho(n,k)^{n-k}.$$

Furthermore, easy computations show that the maximal volume of an $n$-simplex in $B^n$ is $(n+1)^{(n+1)/2}/n^{n/2}n!$ and

$$\mathrm{vol}_n B^n = \frac{\pi^{n/2}}{\Gamma(n/2 + 1)} \leq (2\pi e/n)^n.$$

Putting together the last four relations and the definition of $\rho(n,k)$ we get

$$\mathrm{vol}_n K \leq \binom{m}{k+1} \frac{(k+1)^{(k+1)/2}}{k^{k/2}k!} \frac{\pi^{(n-k)/2}}{\Gamma((n-k+2)/2)} \left( \frac{n-k}{nk} \right)^{(n-k)/2}.$$

Therefore

$$\mathrm{vol}_n K \leq \left( \frac{em}{k+1} \right)^{k+1} \left( \frac{e}{k} \right)^k \left( \frac{2e\pi}{nk} \right)^{(n-k)/2}$$

and so

$$\frac{\mathrm{vol}_n K}{\mathrm{vol}_n B^n} \leq \left( \frac{em}{k+1} \right)^{k+1} n^{k/2} k^{-(n+k)/2}. \qquad (2.3)$$

All that remains is to find a value of $k$ for which the right-hand side is small.

Let's do this under the assumptions $m/n \to \infty$ and $n/\log(m/n) \to \infty$; the existence of $c$ can be shown similarly. We claim that $k = \lceil n/2\log(m/n) \rceil$ is a suitable choice. To avoid too much clutter, we shall just take $k = n/2 \log(m/n)$. Note that $k \to \infty$, $k = o(n)$ and

$$\left(\frac{m}{k+1}\right)^k \le \exp(k \log(m/k)) = \exp\left(k(\log(m/n) + \log(n/k))\right)$$

$$= \exp\left(\frac{n}{2} + n \frac{\log(2 \log(m/n))}{2 \log(m/n)}\right) = e^{(1+o(1))n/2}. \tag{2.4}$$

Also,

$$(n/k)^k = (2 \log(m/n))^k = \exp\left(\frac{n}{2 \log(m/n)} \log(2 \log(m/n))\right) = e^{o(n)},$$

since $m/n \to \infty$. Hence

$$n^{k/2} k^{-(n+k)/2} = (n/k)^{(n+k)/2} n^{-n/2} = e^{o(n)} (2 \log(m/n))^{n/2} n^{-n/2}.$$

Together with (2.3) and (2.4), this implies that

$$\frac{\mathrm{vol}_n K}{\mathrm{vol}_n B^n} \le e^{o(n)} \left(2e \log(m/n)/n\right)^{n/2},$$

completing the proof. □

Theorem 2.5 is essentially best possible in a large range of $m$, except for the constant $2e$. In fact, the theorem can be read out of some earlier results of Carl [1985], and over the years it has been discovered many times, having been published in [Bárány and Füredi 1988; Carl and Pajor 1988; Gluskin 1988; Bourgain, Lindenstrauss, and Milman 1989]. Numerous related results can be found in [Vaaler 1979; Figiel and Johnson 1980; Bárány and Füredi 1986; 1987; Ball and Pajor 1990; Gordon, Reisner, and Schütt $\ge$ 1997], and elsewhere.

We see from Theorem 2.5 that a polytope $K$ contained in $B^n$ with $\mathrm{vol}_n K \ge \frac{1}{2} \mathrm{vol}_n B^n$, say, has exponentially many vertices. In fact, if $\mathrm{vol}_n K$ can be close to $\mathrm{vol}_n B^n$ then $1 - \mathrm{vol}_n K/\mathrm{vol}_n B^n$ is a more significant measure of the volume approximation. Gordon, Reisner and Schütt [$\ge$ 1997] proved that in order to get $1 - \varepsilon$ proportion of the volume, we need about $n^{n/2}$ points: there are positive constants $\varepsilon_0$ and $\varepsilon_1$ such that $1 - \mathrm{vol}_n K/\mathrm{vol}_n B^n \ge \varepsilon_0$ whenever $K$ is a polytope in $B^n$ with $m \le (\varepsilon_1 n)^{n/2}$ vertices—in other words,

$$W(n, m) \le 1 - \varepsilon_0$$

if $m \le (\varepsilon_1 n)^{n/2}$. Even more, there are positive constants $\delta_0$ and $\delta_1$ such that

$$1 - W(n, m) \ge \delta_0 n m^{-2/(n-1)}$$

whenever $n \ge 2$ and $m \ge (\delta_1 n)^{n/2}$.

Many of the papers fleetingly mentioned above concern the volume of the intersection of slabs, rather than the volume of the convex hull of points. In order to prove the main result of this section, namely that computing the volume is difficult, we need one of these results as well. Given natural numbers $n$ and

$m$, let $S(n, m)$ be the infimum of the volumes of intersections of $m$ slabs in $\mathbb{R}^n$, each of the form

$$\{x : |\langle x, v \rangle| \leq 1\},$$

where $v \in \mathbb{R}^n$ is a vector of length at most 1. The intersection of $m$ such slabs is precisely a centrally symmetric polytope $K$ containing $B^n$, with at most $2m$ facets [Ball 1997, Theorem 2.1]. The following lower bound for $S(n, m)$ is given in [Carl and Pajor 1988; Gluskin 1988].

THEOREM 2.6. *There is a constant $\delta > 0$ such that if $1 \leq n \leq m$ then*

$$S(n, m) \geq \left( \frac{\delta}{(\log(m/n) + 1)^{1/2}} \right)^n.$$

Rather than proving this directly (which would not be difficult), we shall deduce it from Theorem 2.5 and a beautiful and important result, the reverse Santaló inequality of Bourgain and Milman. For a convex body $K$ in $\mathbb{R}^n$, the *polar* of $K$ is

$$K^\circ = \{x \in \mathbb{R}^n : \langle x, y \rangle \leq 1 \text{ for all } y \in K\}.$$

If $K$ is a *ball*, that is the unit ball $B(X)$ of a normed space $X = (\mathbb{R}^n, \|\cdot\|)$, then $K^\circ$ is precisely the unit ball of the dual: $K^\circ = B(X^*)$. For us this is precisely the most important case.

What can one say about the product $\operatorname{vol} K \operatorname{vol} K^\circ$ for a ball $K \subset \mathbb{R}^n$? Santaló [1949] proved that it is at most $(\operatorname{vol} B^n)^2$, so that the maximum is attained for $K = B^n$. Thus

$$\operatorname{vol} K \operatorname{vol} K^\circ \leq \pi^n / \Gamma(n/2 + 1)^2 \sim \left( \frac{2\pi e}{n} \right)^n \Big/ \pi n = (2e)^n \pi^{n-1} / n^{n+1}.$$

Taking $X = l_1^n$, so that $X^* = l_\infty^n$, we see that $\operatorname{vol} K \operatorname{vol} K^\circ$ can be as small as $\operatorname{vol} B(l_1^n) \operatorname{vol} B(l_\infty^n) = 4^n / n! \sim (4e/n)^n / \sqrt{2\pi n}$. Mahler conjectured that this value is, in fact, the minimum of the product. Although this long-standing conjecture is still open, Bourgain and Milman [1985] proved the following reverse Santaló inequality, which is only a little weaker than Mahler's conjecture.

THEOREM 2.7. *There is a constant $c_0 > 0$ such that if $K$ is any ball in $\mathbb{R}^n$ then*

$$\operatorname{vol} K \operatorname{vol} K^\circ \geq (c_0/n)^n. \qquad \square$$

It is frequently convenient to state both inequalities together as follows: if $K$ is a ball in $\mathbb{R}^n$ then

$$c_0 \leq \left( \frac{(\operatorname{vol} K)(\operatorname{vol} K^\circ)}{(\operatorname{vol} B^n)^2} \right)^{1/n} \leq 1$$

for some constant $c_0 > 0$.

Theorem 2.6 is an easy consequence of Theorems 2.5 and 2.7. Indeed, let

$$L = \{x \in \mathbb{R}^n : |\langle x, u_i \rangle| \leq 1, \ i = 1, \ldots, m\},$$

where $u_1, \ldots, u_m \in B^n$, be the intersection of $m$ slabs containing $B^n$. Then $L = K^\circ$, where $K = \text{conv}\{u_1, \ldots, u_m\}$. Hence, by Theorems 4 and 5,

$$\text{vol } L \geq (c_0/n)^n / \text{vol } K \geq (c_0/n)^n (n/(\gamma \log(m/n) + 1)^{1/2})^n$$
$$= \left(c_0^2 \gamma (\log(m/n) + 1)\right)^{n/2}.$$

PROOF OF THEOREM 2.1. It suffices to prove the theorem for large $n$. Playing the role of the oracle, we shall give away much more than we have to. First of all, we specify that

$$B(l_1^n) \subset K \subset B(l_\infty^n).$$

Thus $r \leq 1/\sqrt{n}$ and $R \geq \sqrt{n}$ will do, so we can have input size at most $2n$. For $x \in \mathbb{R}^n$, $x \neq 0$, define $x^\circ = x/\|x\|$, $H^+(x^\circ) = \{z \in \mathbb{R}^n : \langle z, x^\circ \rangle \leq 1\}$, and $H^-(x^\circ) = \{z \in \mathbb{R}^n : \langle z, -x^\circ \rangle \leq 1\}$. Here $H^+(x_\circ)$ and $H^-(x_\circ)$ are half-spaces containing $B^n$, and their intersection is a slab.

To the question posed by the algorithm "And what about $x$?", the oracle replies very generously that $x^\circ \in K$, $-x^\circ \in K$, and $K$ is contained in the slab $H^+(x^\circ) \cap H^-(x^\circ)$. This is, of course, consistent with $K = B^n$.

Now let's run the algorithm until $m \leq d^a/2 - n$ questions have been asked for some $a \geq 2$, say $x_1, x_2, \ldots, x_m$. Setting $C = \text{conv}\{\pm e_1, \pm e_2, \ldots, \pm e_n, \pm x_1^\circ, \pm x_2^\circ, \ldots, \pm x_m^\circ\}$, we see that the answers are consistent with $K = C$ and $K = C^\circ$ as well. Consequently,

$$\overline{\text{vol }} C = \overline{\text{vol }} C^\circ \geq \text{vol } C^\circ$$

and

$$\underline{\text{vol }} C \leq \text{vol } C.$$

Therefore

$$\frac{\overline{\text{vol }} C}{\underline{\text{vol }} C} \geq \frac{\text{vol } C^\circ}{\text{vol } C} \geq \frac{S(n, n^a)}{V(n, n^a)}.$$

By Theorems 2.5 and 2.6,

$$\frac{\overline{\text{vol }} C}{\underline{\text{vol }} C} \geq \frac{(\delta/(a \log n)^{1/2})^n}{(\gamma(a \log n)^{1/2}/n)^n} = \left(\frac{n}{\gamma \delta a \log n}\right)^n,$$

proving the assertion. $\qquad\qquad\square$

Numerous related results concerning the hardness of approximations can be found in [Khachiyan 1988; 1989; 1993; Lawrence 1991; Lovász and Simonovits 1992].

To conclude this section, let's say a few words about the volumes of intersections of slabs. For $u_1, \ldots, u_m \in \mathbb{R}^n$, set

$$S(u_1, \ldots, u_m) = \{x \in \mathbb{R}^n : |\langle x, u_i \rangle| \leq 1, \ i = 1, \ldots, m\}.$$

Then Theorem 2.6 claims that if $\max |u_i| \leq 1$ then

$$\text{vol } S(u_1, \ldots, u_m) \geq \{\delta/(\log(m/n) + 1)\}^n.$$

The earliest significant slab-intersection theorem is in [Vaaler 1979], and says that if $\sum_{i=1}^{m} |u_i|^2 \leq n$ then

$$\text{vol } S(u_1, \ldots, u_m) \geq 2^n. \tag{2.5}$$

An attractive reformulation of this result is the following: for $1 \leq k \leq n$, any central section of the unit cube $[-1/2, -1/2]^n$ by a $k$-dimensional subspace has volume at least 1. For $k = n - 1$ this was first proved by Hensley [1979], who also showed that such an $(n-1)$-dimensional intersection has volume at most 5. Subsequently, Ball [1986] improved the upper bound to the following surprising and beautiful best possible result: any section of the unit cube $[-1/2, 1/2]^n$ by an $(n-1)$-dimensional affine subspace has volume at most $\sqrt{2}$.

Clearly, Theorem 2.6 and Vaaler's inequality (2.5) are the $p = \infty$ and $p = 2$ members of a family of inequalities parameterized by $p$. The general case was proved in [Ball and Pajor 1990]: if $1 \leq p < \infty$, $m \geq n$, and $u_1, \ldots, u_m \in \mathbb{R}^n$ are such that $\sum_{i=1}^{m} |u_i|^p \leq r^p n$, then

$$\text{vol } S(u_1, \ldots, u_m) \geq \begin{cases} (2\sqrt{2}/\sqrt{p}\, r)^n & \text{if } p \geq 2, \\ r^{-n} & \text{if } 1 \leq p \leq 2. \end{cases}$$

Vaaler's theorem is the case $p = 2$ of this result. It is interesting to note that Vaaler's theorem was used by Bombieri and Vaaler [Bombieri and Vaaler 1983] to sharpen an important result in the geometry of numbers, namely Siegel's lemma. In turn, Ball and Pajor made use of their extension above to prove a generalization of Siegel's lemma.

## Lecture 3. Rapidly Mixing Random Walks

We saw in the preceding lecture that there is a polynomial-time algorithm that, for every convex body $k \subset \mathbb{R}^n$, produces volume estimates $\underline{\text{vol}}\, K$ and $\overline{\text{vol}}\, K$ satisfying $\overline{\text{vol}}\, K/\underline{\text{vol}}\, K \leq n^n$, and this is the best one can do, except for a factor $(c \log n)^{-n}$. The exciting part of the story is that if we are willing to replace *certainty* by *high probability*—that is, if we are willing to consider *randomized algorithms* that fail with a small probability—then we can do much better. Estimating the volume of a convex body $K$ is akin to sampling at random from the uniform distribution on $K$. In order to find a random point of $K$, one runs a random walk on $K$ (to be precise, a discrete version of $K$) till the distribution of the last point is close to the stationary distribution, which is the uniform distribution. The problem is then to decide when we can stop so that we are likely to be close to the stationary distribution. This leads us to the question of *mixing time*, the time it takes to get close to the stationary distribution, and to criteria for *rapid mixing*, that is getting close to the stationary distribution in unexpectedly few steps. The aim of this section is to give a beautiful and simple condition for rapid mixing in terms of the conductance of the random walk.

Alon and Milman [Alon and Milman 1985; Alon 1986] were the first to connect combinatorial properties—especially expansion properties—of a graph with the second eigenvalue of its Laplace operator. Loosely speaking, a graph $G$ with $n$ vertices *expands well* if, for every set $U$ of at most $n/2$ vertices, there are relatively many edges with precisely one endvertex in $U$. The *Laplacian* of a simple graph $G = (V, E)$ (where $V$ is the set of vertices, $E$ is the set of edges, and *simple* means there is at most one edge joining two vertices and no loops from a vertex to itself) is the linear operator map $Q : L^2(V) \to L^2(V)$ given by the matrix

$$\mathrm{diag}(d(v))_{v \in V} - A,$$

where $d(v)$ is the *degree* of the vertex $v \in V$ (the number of edges incident on $v$) and $A$ is the *adjacency matrix* of $G$ (the matrix whose rows and columns are indexed by $V$ and where each entry is 1 or 0, depending on whether or not there is an edge in $E$ connecting the two vertices in question). (See [Bollobás 1979] for details and other standard terminology.)

Alon and Milman also proved a discrete version of Cheeger's inequality [1970] related to isoperimetric inequalities on manifolds. Connecting expansion with mixing time, Aldous [1987] showed that random walks on graphs with good expansion properties of low degree mix rapidly. Building on these ideas, Jerrum and Sinclair [1989; Sinclair and Jerrum 1989] defined the conductance of a random walk, and showed that large conductance implies fast mixing rate.

Our aim here is to present the connection between conductance and mixing rate. Rather than consider general Markov chains, we shall take essentially the simplest case, that of simple random walks on regular graphs. As so often, it takes no effort to step up from here to a more general setting. We shall follow the simple and elegant approach of [Mihail 1989]; for a more substantial review of random walks, conductances and eigenvalues, see [Vazirani 1991]; for the related spectral properties of graphs, see [Chung 1996].

Let $G = (V, E)$ be a connected $d$-regular simple graph ($d$-regular means that every vertex has degree $d$). We write $V = \{1, \ldots, n\}$ for notational simplicity. For the purposes of these lectures, a *simple random walk on $G$ with initial state $X_0$* is a sequence of random variables

$$\tilde{X} = (X_0, X_1, \ldots),$$

taking values in $V$, such that for $i, j \in V$ and $t \geq 0$ we have

$$\mathbb{P}(X_{t+1} = j \mid X_t = i) = \begin{cases} \frac{1}{2} & \text{if } i = j, \\ 1/2d & \text{if } ij \in E, \\ 0 & \text{otherwise.} \end{cases}$$

Intuitively, if $X_t$ represents the probability distribution of the random walker's position at time $t$; that is, $\mathbb{P}(X_t = i)$ is the probability that she will be at vertex $i$ at time $t$. The display above says that from time $t$ to time $t + 1$ the random walker has a 50% chance of staying put, and equal chances of moving away from the current vertex along any of the edges incident on it. Since the transition probabilities are independent of $t$, the sequence $(X_0, X_1, \ldots)$ is a Markov chain.

The elementary theory of Markov processes guarantees that, with these probabilities, $\mathbb{P}(X_t = i) \to 1/n$ for all $i$ as $t \to \infty$, no matter what the initial state is (see [Kemeny and Snell 1976], for example). The question is, how fast? Set $p_i^{(t)} = \mathbb{P}(X_t = i)$ and let $e_{i,t} = p_i^{(t)} - 1/n$ be the *excess probability* at $i$. The excess probabilities satisfy

$$e_{i,\,t+1} = p_i^{(t+1)} - \frac{1}{n} = \left( \tfrac{1}{2} p_i^{(t)} + \frac{1}{2d} \sum_{j \in \Gamma(i)} p_j^{(t)} \right) - \frac{1}{n}$$

$$= \tfrac{1}{2} \left( p_i^{(t)} - 1/n \right) + \frac{1}{2d} \sum_{j \in \Gamma(i)} \left( p_j^{(t)} - 1/n \right)$$

$$= \tfrac{1}{2} e_{i,t} + \frac{1}{2d} \sum_{j \in \Gamma(i)} e_{j,t} = \frac{1}{2d} \sum_{j \in \Gamma(i)} (e_{i,t} + e_{j,t}). \qquad (3.1)$$

Define

$$d_1(t) = d_1(\tilde{X}, t) = \sum_i |e_{i,t}|$$

and

$$d_2(t) = d_2(\tilde{X}, t) = \sum_i e_{i,t}^2.$$

A simple random walk $\tilde{X}$ on $G$ is *rapidly mixing* if there is a polynomial $f$ such that if $0 < \varepsilon < \tfrac{1}{3}$ and $t \geq f(\log n) \log(1/\varepsilon)$ then $d_1(t) \leq \varepsilon$.

Strictly speaking, this definition does not make much sense since if we have only one graph $n$ itself is really a constant. For a proper definition, we need a *sequence* $(G_i)_{i=1}^{\infty}$ of regular graphs, where each $G_i$ has $n_i$ vertices and $n_i \to \infty$. We say that the simple random walks on $G_1, G_2, \ldots$ are *rapidly mixing* if there is a polynomial $f$, depending only on the sequence $(G_i)$, such that if $0 < \varepsilon < \tfrac{1}{3}$ and $t \geq f(\log n_i) \log(1/\varepsilon)$ then $d_1(\tilde{X}_i, t) \leq \varepsilon$ whenever $\tilde{X}_i$ is a simple random walk on $G_i$.

Let's define the *conductance* of $G$ or the *conductance* of a simple random walk on $G$ as follows. For $U \subset V$ set $\bar{U} = V - U$ and

$$\Phi_G(U) = \frac{e(U, \bar{U})}{d\,|U|}.$$

Note that $0 \leq \Phi_G(U) \leq 1$. Also, for $1 \leq |U| \leq n/2$, $\Phi_G(U)$ is small if there are relatively few $U - \bar{U}$ edges, that is, if there is a "bottleneck" when we try to go from $U$ to $\bar{U}$. The *conductance* of $G$ is then

$$\Phi_G = \min_{|U| \leq n/2} \Phi_G(U).$$

The conductance is also called the *isoperimetric number* of the graph or its *Cheeger constant*. The quantity $d\,|U|$ is the "volume" of $U$, the sum of the degrees of its vertices. If $G = (V, E)$ is not necessarily regular then for $U \subset V$ the *volume* of $U$ is $\mathrm{vol}\, U = \sum_{u \in U} d(u)$, and the conductance of $G$ is

$$\min_{U \subset V} \frac{e(U, \bar{U})}{\min\{\mathrm{vol}\, U,\, \mathrm{vol}\, \bar{U}\}}.$$

With this definition, the results below are easily extended to general graphs and beyond; we shall state one of these results at the end of the lecture.

Clearly, we have $0 \leq \Phi_G \leq 1$, although the upper bound is somewhat unrealistic: if $\Phi_G = 1$ then $G$ is either the trivial graph consisting of one vertex, or a single edge, or a triangle. If $G$ has many vertices, the best we can hope is that $\Phi_G$ is not far from $\frac{1}{2}$. Concerning the lower bound, note that $\Phi_G = 0$ if and only if $G$ is disconnected.

Our main aim is to prove the following fundamental result, which clearly shows the importance of the conductance.

THEOREM 3.1. *Every simple random walk on $G$ satisfies*

$$d_2(t + 1) \leq \left(1 - \tfrac{1}{4}\Phi_G^2\right) d_2(t).$$

*In particular, as $d_2(0) \leq 2$,*

$$d_2(t) \leq \left(1 - \tfrac{1}{4}\Phi_G^2\right)^t d_2(0) \leq 2\left(1 - \tfrac{1}{4}\Phi_G^2\right)^t.$$

We shall deduce this result from two lemmas that are of interest in their own right.

LEMMA 3.2. $d_2(t + 1) \leq d_2(t) - \dfrac{1}{2d} \displaystyle\sum_{ij \in E} (e_{i,t} - e_{j,t})^2.$

PROOF. By (3.1),

$$d_2(t + 1) = \frac{1}{4d^2} \sum_{i=1}^{n} \left( \sum_{j \in \Gamma(i)} (e_{i,t} + e_{j,t}) \right)^2.$$

Applying the Cauchy–Schwarz inequality to the inner sum, we find that, as $|\Gamma(i)| = d$,

$$d_2(t + 1) \leq \frac{1}{4d^2} \sum_{i=1}^{n} \left( \sum_{j \in \Gamma(i)} (e_{i,t} + e_{j,t})^2 \right) d$$

$$= \frac{1}{2d} \sum_{ij \in E} (e_{i,t} + e_{j,t})^2 = \frac{1}{2d} \sum_{ij \in E} \left\{ 2\left(e_{i,t}^2 + e_{j,t}^2\right) - (e_{i,t} - e_{j,t})^2 \right\}$$

$$= d_2(t) - \frac{1}{2d} \sum_{ij \in E} (e_{i,t} - e_{j,t})^2. \qquad \square$$

The second lemma needs a little more work.

LEMMA 3.3. *Suppose weights $x_i$ are assigned to the elements of the vertex set $V = \{1, \ldots, n\}$, satisfying $\sum_{i=1}^{n} x_i = 0$. Then*

$$\sum_{ij \in E} (x_i - x_j)^2 \geq \frac{d}{2} \Phi_G^2 \sum_{i=1}^{n} x_i^2.$$

PROOF. Set $m = \lceil n/2 \rceil$. We shall prove that if $y_1 \geq y_2 \geq \ldots \geq y_n$, with $y_m = 0$, then

$$\sum_{ij \in E} (y_i - y_j)^2 \geq \frac{d}{2} \Phi_G^2 \sum_{i=1}^{n} y_i^2. \tag{3.2}$$

This is stronger than the desired inequality. Indeed, in the statement of the lemma we may assume that $x_1 \geq x_2 \geq \ldots \geq x_n$. Setting $y_i = x_i - x_m$, inequality (3.2) gives

$$\sum_{ij \in E} (x_i - x_j)^2 = \sum_{ij \in E} (y_i - y_j)^2 \geq \frac{d}{2} \Phi_G^2 \sum_{i=1}^{n} (x_i - x_m)^2 = \frac{d}{2} \Phi_G^2 \sum_{i=1}^{n} x_i^2 + \frac{nd}{2} \Phi_G^2 x_m^2,$$

since $\sum_{i=1}^{n} x_i = 0$.

In order to prove (3.2), set

$$u_i = \begin{cases} y_i & \text{if } i \leq m, \\ 0 & \text{if } i > m, \end{cases} \qquad v_i = \begin{cases} 0 & \text{if } i \leq m, \\ y_i & \text{if } i > m. \end{cases}$$

Thus $y_i = u_i + v_i$ for every $i$. Also, if $u_i \neq 0$ then $u_i > 0$ and $i < m$, and if $v_i \neq 0$ then $v_i < 0$ and $i > m$. Since $(y_i - y_j)^2 = (u_i - u_j + v_i - v_j)^2 \geq (u_i - u_j)^2 + (v_i - v_j)^2$ for every edge $ij$, it suffices to prove that

$$\sum_{ij \in E} (u_i - u_j)^2 \geq \frac{d}{2} \Phi_G^2 \sum_{i=1}^{m} u_i^2 \tag{3.3}$$

and

$$\sum_{ij \in E} (v_i - v_j)^2 \geq \frac{d}{2} \Phi_G^2 \sum_{i=m}^{n} v_i^2.$$

Furthermore, as $m \geq n - m$, it suffices to prove (3.3). We may assume that $u_1 > 0$. By the Cauchy–Schwarz inequality,

$$\left( \sum_{ij \in E} (u_i^2 - u_j^2) \right)^2 = \left( \sum_{ij \in E} (u_i - u_j)(u_i + u_j) \right)^2 \leq \sum_{ij \in E} (u_i - u_j)^2 \sum_{ij \in E} (u_i + u_j)^2$$

$$\leq \sum_{ij \in E} (u_i - u_j)^2 \sum_{ij \in E} 2 \left( u_i^2 + u_j^2 \right)$$

$$= 2d \sum_{i=1}^{n} u_i^2 \sum_{ij \in E} (u_i - u_j)^2. \tag{3.4}$$

We may assume that, in all the sums $\sum_{ij \in E}$ over the edges $ij$, we have $i < j$. Note that

$$\sum_{ij \in E} (u_i^2 - u_j^2) = \sum_{ij \in E} \sum_{l=i}^{j-1} (u_l^2 - u_{l+1}^2) = \sum_{l=1}^{n-1} (u_l^2 - u_{l+1}^2) \, e(U_l, \bar{U}_l),$$

where $U_l = \{1, \ldots, l\}$ and $\bar{U}_l = \{l+1, \ldots, n\}$. Since $u_m = u_{m+1} = \ldots = u_n = 0$, this gives

$$\sum_{ij \in E} (u_i^2 - u_j^2) = \sum_{l=1}^{m-1} (u_l^2 - u_{l+1}^2) \, e(U_l, \bar{U}_l) \geq \sum_{l=1}^{m-1} (u_l^2 - u_{l+1}^2) \, d\,\Phi_G l$$

$$= d\Phi_G \sum_{l=1}^{m-1} u_l^2 = d\Phi_G \sum_{l=1}^{n} u_l^2. \tag{3.5}$$

Inequalities (3.4) and (3.5) give

$$\sum_{ij \in E} (u_i - u_j)^2 \geq \left( d\Phi_G \sum_{i=1}^{n} u_i^2 \right)^2 \bigg/ \left( 2d \sum_{i=1}^{n} u_i^2 \right) = \frac{d}{2} \Phi_G^2 \sum_{i=1}^{n} u_i^2,$$

as desired.  □

PROOF OF THEOREM 3.1. By Lemma 3.2,

$$d_2(t) - d_2(t+1) \geq \frac{1}{2d} \sum_{ij \in E} (e_{i,t} - e_{j,t})^2.$$

Applying Lemma 3.3 with $x_i = e_{i,t}$, we find that

$$d_2(t) - d_2(t+1) \geq \tfrac{1}{4}\Phi_G^2 \sum_{i=1}^{n} e_{i,t}^2 = \tfrac{1}{4}\Phi_G^2 d_2(t),$$

completing the proof.  □

By the Cauchy–Schwarz inequality, $d_1(t) \leq (n\,d_2(t))^{1/2}$, so Theorem 3.1 has the following immediate consequence.

COROLLARY 3.4. *Every simple random walk on a graph $G$ of order $n$ with conductance $\Phi_G$ satisfies*

$$d_1(t) \leq (2n)^{1/2} \left(1 - \tfrac{1}{4}\Phi_G^2\right)^{t/2}.  \qquad \square$$

Corollary 3.4 implies that if we assume that $G$ connected (so that $\Phi_G > 0$), that $0 < \varepsilon < 1/3$, and that

$$t > 8\Phi_G^{-2} \left( \log(1/\varepsilon) + \tfrac{1}{2}\log(2n) \right)$$

then

$$d_1(t) \leq (2n)^{1/2} \left(1 - \tfrac{1}{4}\Phi_G^2\right)^{t/2} < \exp\left(\tfrac{1}{2}\log(2n) - \tfrac{1}{8}\Phi_G^2 t\right) < \varepsilon.$$

In particular, if $n \geq 3$ and $t \geq 8\Phi_G^{-2} \log n \log(1/\varepsilon)$ then $d_1(t) < \varepsilon$. This gives us the following sufficient condition for rapid mixing.

THEOREM 3.5. *Let $(G_i)_1^\infty$ be a sequence of regular graphs with $|G_i| = n_i \to \infty$. If there is a $k \in \mathbb{N}$ such that*

$$\Phi_{G_i} \geq (\log n_i)^{-k}$$

*for sufficiently large $i$, the simple random walks on $(G_i)_1^\infty$ are rapidly mixing.*

PROOF. We have just seen that $f(x) = 8x^{2k+1}$ will do if $n_i \geq 3$. □

Let's see some families of regular graphs for which we can give a good lower bound for the conductance. As a trivial example, take the complete graph $K_n$. It is immediate that $\Phi_{K_n} > \frac{1}{2}$ for $n \geq 2$, so the simple random walks on $(K_n)$ are rapidly mixing. Of course, this can be derived very simply from first principles as well.

As a less trivial example, we take the cubes $Q_1, Q_2, \ldots$, defined as follows: the vertex set of $Q_d$ is $\{0,1\}^d$, the set of sequences $x = (x_i)_1^n$, $x_i = 0$ or 1, and two sequences joined by an edge if they differ in only one term. $Q_d$ is obviously $d$-regular, and it is easy to prove that $\Phi_{Q_d} = 1/d$. The worst bottlenecks arise between the "top" and "bottom" of $Q_n$: for $U = \{(x_i) \in Q_d : x_1 = 1\}$ and $\bar{U} = \{(x_i) \in Q_d : x_1 = 0\}$, say. Clearly, $e(U, \bar{U}) = |U| = 2^{d-1}$ so that $\Phi_{Q_n}(U) = 1/d$.

Since $\Phi_{Q_d} = 1/d = 1/\log n$, where $n = 2^d = |Q_d|$, simple random walks on $(Q_d)_1^\infty$ are rapidly mixing.

The cube $Q_d$ is just $K_2^d = K_2 \times \ldots \times K_2$, that is, the product of $d$ paths of lengths 1. Taking the product of $d$ cycles, each of length $l$, we get the torus $T_l^d$. This graph has $l^d$ vertices and it is $2d$-regular. One can show that for $G = T_{2l}^d$ we have $\Phi_G = 2/(ld)$. (Note that $T_4^d$ is just the cube $Q_{2d}$.) Hence, for a fixed value of $l$, simple random walks on $(T_{2l}^d)_{d=1}^\infty$ are rapidly mixing.

It is straightforward to extend Theorem 3.1 to aperiodic random walks. To be precise, let $V$ be a finite set and let $X$ be a random walk on $V$ with transition probabilities $p(u,v)$ such that $p(u,u) \geq \frac{1}{2}$. Suppose that $X$ is *reversible*, that is, there is a (stationary) probability distribution $\lambda$ on $V$ with $\lambda(u)p(u,v) = \lambda(v)p(v,u)$. Here it is natural to view $\lambda(u)p(u,v)$ as the *flow* from $u$ to $v$: it is the same as the flow from $v$ to $u$. Such a random walk is a straightforward generalization of a simple random walk on a regular graph discussed above; in fact, it is hardly more than that. Denoting by $\lambda(U) = \sum_{u \in U} \lambda(u)$ the "volume" of a $U \subset V$, the *conductance* of $X$ is

$$\tilde{\Phi}_X = \min_{\lambda(U) \leq 1/2} \frac{\sum_{u \in U} \sum_{v \in V \setminus U} \lambda(u)p(u,v)}{\lambda(U)}.$$

Note that this definition makes the conductance half as large as before since if $X$ is the simple random walk on a $d$-regular graph then $p(u,v) = 1/(2d)$, so $\tilde{\Phi}_X = \frac{1}{2}\Phi_G$. The fact is that for random walks this is the natural definition, while for graphs $\Phi_G$ is natural.

Needless to say, we are interested in convergence to the stationary distribution $\lambda$. To measure the distance from $\lambda$, as before, we put

$$d_2(t) = \sum_{v \in V} \left(p_v^{(t)} - \lambda(v)\right)^2.$$

Let's state then the analogue of Theorem 3.1: the proof is unchanged.

THEOREM 3.6. *With the notation above,*

$$d_2(t+1) \le \left(1 - \tilde{\Phi}_X^2\right) d_2(t)$$

*and so*

$$d_2(t) \le 2\left(1 - \tilde{\Phi}_X^2\right)^t. \qquad \Box$$

This is the theorem we shall need in the next section.

## Lecture 4. Randomized Volume Algorithms

We have seen that no polynomial-time algorithm can estimate the volume of a convex body substantially better than within a factor of $n^n$. Thus, if we want our algorithm to produce a lower bound and an upper bound that are guaranteed to be valid in *every instance* and be reasonably fast, we cannot demand that the ratio of the two bounds be substantially less than $n^n$. The situation is entirely different if we allow *randomization* and do not insist that the bounds of the algorithm be valid every time, only that they be valid with *high probability*.

Estimating the volume can be viewed as a game between *Hider*, trying to "hide" the volume of a convex body, and *Seeker*, the algorithm trying to pin down the volume. In the case of a deterministic algorithm, Hider is allowed to change his mind as the game progresses: to be precise, there is no way of telling whether he changes his mind or not, as all he has to make sure is that the answers he gives remain consistent with *some* convex body. On the other hand, a randomized algorithm is applicable only if Hider is required to play an honest game, that is if he has to fix a convex body once and for all at the beginning of the game. Then Seeker may keep tossing coins in order to decide his next appeal to the oracle, and so he may come up with a randomized algorithm that gets good results fast, with probability close to 1. Seeker trades certainty for speed and efficiency, with large probability. The probability of failure should be small and independent of the body Hider chooses.

Let's assume that our body $K \subset \mathbb{R}^n$, where $n \ge 2$, is given by a well-guaranteed strong membership oracle (although, as we mentioned earlier, it is unimportant which membership oracle we take). Let $\varepsilon$ and $\eta$ be small positive numbers, say less than $\frac{1}{3}$. An *$\varepsilon$-approximation* to $\mathrm{vol}\,K$ is a number $\widetilde{\mathrm{vol}}\,K$ such that

$$(1 - \varepsilon)\widetilde{\mathrm{vol}}\,K < \mathrm{vol}\,K < (1 + \varepsilon)\,\mathrm{vol}\,K.$$

If all goes well, we may hope to find a *fully polynomial approximation scheme* (FPRAS) for approximating the volume of a convex body: a randomized algorithm that runs in time polynomial in $\langle K \rangle$, $1/\varepsilon$ and $\log(1/\eta)$, and with probability at least $1 - \eta$ produces an $\varepsilon$-approximation to $\mathrm{vol}\, K$.

In 1989, Dyer, Frieze, and Kannan [1991] found precisely such an algorithm. In describing the speed of an FPRAS, it is convenient to use the "soft-$O$" notation $O^*$, one that ignores powers of $\log n$ and polynomials of $1/\varepsilon$ and $\log(1/\eta)$. In this notation, Dyer, Frieze, and Kannan produced an FPRAS running in time $O^*(n^{23})$—to be precise, in time $O(n^{23}(\log n)^5 \varepsilon^{-2}(\log 1\varepsilon)(\log 1/\eta))$. With this result, the floodgates opened: Lovász and Simonovits [1990] found a $O^*(n^{16})$ algorithm, Applegate and Kannan [1991] and Lovász and Simonovits [1992] reduced the complexity to $O^*(n^{10})$, then Dyer and Frieze [1991] to $O^*(n^8)$, Lovász and Simonovits [1993] to $O^*(n^7)$, and Kannan, Lovász, and Simonovits [$\geq$ 1997] to $O^*(n^5)$.

All algorithms are modelled on the original algorithm of Dyer, Frieze and Kannan, so they use a multiphase Monte Carlo algorithm to reduce volume computation to sampling, and use random walks to sample. In order to decide when we are likely to be close to the stationary distribution, conductance is used to bound the mixing time. Finally, isoperimetric inequalities are used to bound the conductance.

In this lecture, we shall sketch one of the most beautiful of these algorithms, given in [Dyer and Frieze 1988]. We do not give nearly all the details, to avoid making the presentation too technical.

Before stating the result, let's say a few words about an obvious naive approach to estimating the volume by a randomized algorithm, which goes as follows. Place a fine grid on $K$: for example, assuming $n^2 B_\infty^n \subset K \subset n^4 B_\infty^n$, we may take $\mathbb{Z}^n \cap n^4 B_\infty^n$. Consider a random walk on this grid, whose stationary distribution is exactly the uniform distribution on the grid. Run such a random walk long enough so that it gets close to the stationary distribution. Stop it and check whether the point is in $K$ or not. Roughly, with probability $\mathrm{vol}\, K/(2n^4)^n$ we should get a point of $K$, so running this walk sufficiently many times, we should be able to estimate $\mathrm{vol}\, K$.

All this, of course, leads nowhere, since the probability of our random walk ending in $K$ is likely to be exponentially small: it need not even be more than $n^{-n}$. Thus, to estimate it, we would have to run $O(n^n)$ walks.

The moral of all this is that we should try to estimate only rather large *ratios* of volumes. The problem is easily reduced to this, but at the expense of not knowing the shape of the larger body either. Thus what we can have is two bodies $L \subset K$, given by oracles, with $\mathrm{vol}\, L > \frac{1}{2}\mathrm{vol}\, K$, say, and our task is to estimate $\mathrm{vol}\, L/\mathrm{vol}\, K$. Now it would be good enough to estimate this ratio by running a random walk on the part of a fine grid inside $K$, with the uniform distribution being the stationary distribution. But now the problem is that we would like to define a random walk on the grid inside a body $K$ we know almost nothing about, in a way that makes its stationary distribution uniform. This is a pretty tall order.

The following beautiful idea solves our difficulty. Define on a set *larger* than the grid inside $K$ (an entire grid graph, say) a random walk with the following properties: the stationary distribution is uniform on the points belonging to $K$; the stationary distribution gives a fairly large probability to the set of points in $K$; and the walk converges to the stationary distribution fast: it is rapidly mixing. At the first sight, all this seems to be a pie in the sky, but the beauty of it all is that Dyer, Frieze, and Kannan managed to define precisely such a random walk.

Needless to say, there are a good number of technical difficulties to overcome, the most important of which is that the random walk is rapidly mixing. This is proved with the aid of isoperimetric inequalities.

After this preamble, let's state the main result of this lecture, first proved in [Dyer, Frieze, and Kannan 1991], and sketch its proof.

THEOREM 4.1. *There is a fully polynomial randomized approximation scheme for the volume of a convex body given by a well-guaranteed membership oracle.*

PROOF. Sketch of proof We divide the proof into seven steps, saying rather little about each. Let $K \subset \mathbb{R}^n$ be a convex body given by the strong membership oracle, with guarantees $r \geq 1$ and $R$, so that the size of the input is $\langle K \rangle = n + \lceil \log_2 r \rceil + \lceil \log_2 R \rceil$. As always, we may assume that $n$ is large.

1. **Rounding.** Let's write $B_\infty^n = \{x \in \mathbb{R}^n : |x_i| \leq 1 \text{ for every } i\}$ for the cube of side length 2 centred at the origin. There is a polynomial algorithm that replaces $K$ by its affine image (also denoted by $K$) such that

$$2n^2 B_\infty^n \subset K \subset 2n^4 B_\infty^n,$$

say. The ratio $n^2$ of the radii of the balls is rather unimportant: $n^{20}$ would do just as well. In fact, one can do much better: making use of an idea of Lenstra [1983], Applegate and Kannan [1991] showed that we can achieve

$$B_\infty^n \subset K \subset 2(n+1)B_\infty^n$$

as well. To this end, we start with a right simplex $S$ in $K$ and gradually expand it. By rescaling everything so that $S$ becomes the standard simplex $\mathrm{conv}\{0, e_1, e_2, \ldots, e_n\}$, where $(e_i)_1^n$ is the standard basis of $\mathbb{R}^n$, one can check in polynomial time whether the region $\{x \in K : |x_i| \geq 1 + 1/n^2\}$ is empty. If it is not empty, we replace $S$ by a simplex $S' \subset K$ with $\mathrm{vol}\, S' \geq (1 + 1/n^2)\,\mathrm{vol}\, S$, and if it is empty we terminate the process.

**2. Subdivision.** Let's place a number of cubes $C_i = r_i B_\infty^n$ between the cube $C_0 = 2n^2 B_\infty^n$ and $C_l = 2n^4 B_\infty^n$, where $l = \lceil 2(n+1)\log_2 n \rceil$: we take $r_i = \lfloor 2^{i/(n+1)} n^2 \rfloor$ for $0 \le i < l$. Also, set $K_i = C_i \cap K$, so that $K_0 = C_0$ and $K_l = K$. With

$$\alpha_i = \operatorname{vol} K_{i-1} / \operatorname{vol} K_i,$$

we have

$$\operatorname{vol} K = \frac{\operatorname{vol} K_l}{\operatorname{vol} K_{l-1}} \cdot \frac{\operatorname{vol} K_{l-1}}{\operatorname{vol} K_{l-2}} \cdots \cdot \frac{\operatorname{vol} K_1}{\operatorname{vol} K_0} \cdot \operatorname{vol} K_0 = 2^n \Big/ \prod_{i=1}^{l} \alpha_i.$$

Hence it suffices to find an approximation of each $\alpha_i$.

At the first sight, this does not seem to be much of a progress. However, since

$$K_i \subset \frac{r_i}{r_{i-1}} K_{i-1}$$

and

$$r_i/r_{i-1} \le 2^{1/(n+1)} \frac{r_{i-1}+1}{r_{i-1}} < (1+1/2n^2) 2^{1/(n+1)} < 2^{1/n},$$

we have

$$\alpha_i = \frac{\operatorname{vol} K_{i-1}}{\operatorname{vol} K_i} > \frac{1}{2}.$$

Thus the gain is that it suffices to approximate the proportion of the volume of a convex body $L$ in a convex body $K$, when this proportion is rather large. In other words, we do not have to search for an exponentially small body inside another body, only for a body taking up at least half of the volume.

**3. Density.** With a slight abuse of notation, we set $K = K_i$ and $L = K_{i-1}$, where $1 \le i \le l$. Thus

$$K_0 = 2n^2 B_\infty^n \subset L \subset K \subset K_l = wn^4 B_\infty^n$$

and

$$L \subset K \subset 2^{1/n} L.$$

Let $V$ be the set of lattice points $\mathbb{Z}^n$ in $K_l$ and let $G = (V, E)$ be the subgraph of $\mathbb{Z}^n$ induced by $V$. Then $G$ is the *grid graph* $P_m^n$, with $m = 4n^4 + 1$, having $m^n$ vertices: the product of $n$ paths, each of length $4n^4$ and so with $m = 4n^4 + 1$ vertices.

We shall define a distribution on $V$ that will turn out to be the stationary distribution of a certain random walk on the graph $G$.

First, for $x \in \mathbb{R}^n$ set

$$\varphi_0(x) = \min\big\{s \ge 0 : n^2 x \in (n^2 + s) K\big\}.$$

Clearly, $\varphi_0$ is a convex function that varies at most 1 on points at distance at most 1 in $\|\cdot\|_\infty$: if $\|x - y\|_\infty \le 1$ then $n^2(x-y) \in n^2 B_\infty^n \subset K$, so $n^2 x = n^2 y + n^2(x-y) \in (n^2 + \varphi_0(y)) K + K = (n^2 + \varphi_0(y) + 1) K$, giving $\varphi_0(x) \le \varphi_0(y) + 1$.

For $x \in V = \mathbb{Z}^n \cap K_l$, set $\varphi(x) = \lceil \varphi_0(x) \rceil$, and let $\varphi_1(x)$ be the maximal convex function on $K_l$ dominated by $\varphi$. Then

$$\varphi_0(x) \le \varphi_1(x) \le \varphi(x) < \varphi_0(x) + 1$$

for $x \in V$. Finally, for $x \in K_l$, set

$$f(x) = 2^{-\varphi(x)} \qquad \text{and} \qquad f_1(x) = 2^{-\varphi_1(x)}.$$

Then

$$\tfrac{1}{2} f_1(x) < f(x) \le f_1(x)$$

and $f(x) = 1$ on $K$.

Our aim is to define a random walk on the grid graph $G$ whose stationary distribution is precisely $f$, suitably normalized.

**4. The random walk.** Let's define a random walk on the grid graph $G = (V, E)$ by giving the transition probabilities:

$$p(x,y) = \begin{cases} 1/(4n) & \text{if } xy \in E \text{ and } \varphi(y) \le \varphi(x), \\ 1/(8n) & \text{if } xy \in E \text{ and } \varphi(y) = \varphi(x) + 1, \\ 1 - \sum_{z \in \Gamma(x)} p(x,z) & \text{if } y = z, \\ 0 & \text{otherwise.} \end{cases}$$

Thus for every $x \in V$ our random walk stays put at $x$ with probability at least $\frac{1}{2}$; with probability $1/4n$ it goes to a neighbouring vertex $y$ of "norm" $\varphi(y) \le \varphi(x)$, and with *half as much* probability it goes to a neighbour of larger "norm" $\varphi(y)$.

This random walk is *reversible*, with stationary distribution $\lambda(x) = cf(x)$, where $c > 0$ is a normalizing constant. (Thus $c\sum_{x \in V} f(x) = 1$.) Indeed, if $xy \in E$ and $\varphi(x) = \varphi(y)$ then

$$f(x)p(x,y) = 2^{-\varphi(x)} \frac{1}{4n} = 2^{-\varphi(y)} \frac{1}{4n} = f(y)p(y,x),$$

and if $\varphi(x) + 1 = \varphi(y)$ then

$$f(x)p(x,y) = 2^{-\varphi(x)} \frac{1}{8n} = 2^{-\varphi(y)} \frac{1}{4n} = f(y)p(y,x).$$

Another very important aspect of this random walk is that, although it has been tailored for $K$, it is very efficient to compute the transition probabilities at the points *where we need it*. (We certainly cannot afford to compute the transition probabilities at *all* the vertices! That would need exponentially many steps.) All we have to do is to *carry the value of $\varphi$*: as $\varphi$ is known to change by at most one at the next step, at most $4n$ appeals to the oracle give us the values of $\varphi$ at all the neighbours. Having got these values, we know all the transition probabilities from our point, so we can take the next step of our random walk. To keep things simple, we start from a point of $K$, say from $O \in n^2 B_\infty^n \subset K$; then $\varphi(O) = O$, and we are away.

**5. The error term.** For the probability distribution $\lambda$ on $V$, we have

$$\lambda(K) = \lambda(V \cap K) > \frac{1}{2}. \qquad (4.1)$$

In other words, running our random walk long enough, the probability of ending in $K$ is more than $\frac{1}{2}$. Indeed,

$$f(K) = \sum_{x \in K \cap V} f(x) = |K \cap V| \sim \mathrm{vol}\, K.$$

Also, if $x \in \mathbb{Z}^n$ and for a positive integer $s$ we have

$$x \in \left\{ \left(1 + \frac{s}{n^2}\right) K - \left(1 + \frac{s-1}{n^2}\right) K \right\}$$

then $s - 1 < \varphi_0(x) \le \varphi(x) \le s$, so $f(x) = 2^{-s}$. The number of lattice points satisfying (4.1) is about the volume of the body on the right-hand side, so about

$$\left\{ \left(1 + \frac{s}{n^2}\right)^n - \left(1 + \frac{s-1}{n^2}\right)^n \right\} \mathrm{vol}\, K,$$

so it is certainly at most

$$\left(e^{s/n} - 1\right) f(K).$$

Hence,

$$f(\mathbb{Z}^n) = \sum_{x \in \mathbb{Z}^n} f(x) \le f(K) + \sum_{s=1}^{\infty} 2^{-s} \left(e^{s/n} - 1\right) f(K) < 2f(K).$$

But $\lambda$ is just $cf$, so

$$\lambda(K) = \lambda(K)/\lambda(\mathbb{Z}^n) = f(K)/f(\mathbb{Z}^n) > \tfrac{1}{2},$$

proving the claim.

**6. A coin toss.** Set $\alpha = \mathrm{vol}\, L/\mathrm{vol}\, K$ and $\alpha' = \lambda(L \cap V)/\lambda(K \cap V) = \lambda(L)/\lambda(K)$. Then $\alpha'$ is sufficiently close to $\alpha$, so all we have to do is estimate $\alpha'$. This will be done by tossing a biased coin, with probability about $\alpha'$ for heads. Here is how one coin toss works.

We run our random walk long enough, till it is close enough to the stationary distribution. Say, we stop our random walk $X_0 = 0$, $X_1$, ... at $X_t \in \mathbb{Z}^d$. Let $E_0$ be the event that $X_t \notin K$, $E_1$ the event that $X_t \in L$ and $E_2$ the event that $X_t \in K \setminus L$. Then $\mathbb{P}(E_0)$ is not too large: by (4.1), it is not much larger than $\frac{1}{2}$, so $\mathbb{P}(E_1 \cup E_2)$ is substantial, at least $\frac{1}{3}$. Since, on the lattice points of $K$, the stationary distribution $\lambda$ is uniform, we have

$$\mathbb{P}(E_2)/\mathbb{P}(E_1 \cup E_2) \sim \alpha',$$

with good enough approximation.

By repeating the coin toss sufficiently many times, our approximation will be good enough with high enough probability.

**7. The crunch.** What we have to show now is that it suffices to run our random walk for polynomial time to get close to its stationary distribution—in short, that our random walk is rapidly mixing. By Theorem 3.6, all we need is to show that our random walk has large enough conductance. That this is the case, and so Theorem 4.1 holds, follows from the following *isoperimetric inequality*, essentially due to Lovász and Simonovits[Lovász and Simonovits 1990].

THEOREM 4.2. *Let $M \subset \mathbb{R}^n$ be a convex body and let $\mathcal{B}(M)$ be the $\sigma$-field of Borel subsets of $M$. Let $F : \operatorname{Int} M \to \mathbb{R}^+$ be a log-concave function and let $\mu$ be the measure on $\mathcal{B}(M)$ with density $F$:*

$$\mu(A) = \int_A F\,dx$$

*for $A \in \mathcal{B}(M)$. Then, for $A_1, A_2 \in \mathcal{B}(M)$, we have*

$$\min\left\{\mu(A_1), \mu(A_2)\right\} \leq \tfrac{1}{2}\,\frac{\operatorname{diam} M}{d(A_1, A_2)}\,\mu(M \setminus A_1 \cup A_2),$$

*where $\operatorname{diam} M = \max\{|x - y| : x, y \in M\}$ is the diameter of $M$ and $d(A_1, A_2) = \inf\{|x - y| : x \in A_1,\, y \in A_2\}$ is the distance between $A_1$ and $A_2$.* □

The proof uses repeated bisections and is somewhat similar to the method of [Payne and Weinberger 1960]. Lovász and Simonovits [1990] proved the inequality with a constant 1 instead of the best possible constant $\tfrac{1}{2}$, which was inserted in [Dyer and Frieze 1991].

Let's see then that Theorem 4.2 implies that the conductance of our random walk is not too small. Let $U \subset V$, $0 < \lambda(U) < \tfrac{1}{2}$, $\bar{U} = V \setminus U$, and let $\partial U$ be the boundary of $U$, that is, the set of vertices in $\bar{U}$ having at least one neighbour in $U$. Let $M$ be the union of unit cubes centred at the points of $V$, so that $M$ is a solid cube. Let $A_1$ be the union of unit cubes centred at the vertices of $U$, let $B$ be the union of cubes of volume 2 centred at the vertices of $\partial U$, and set $A_2 = M \setminus (A_1 \cup B)$.

Writing $c_1, c_2, \ldots$ for positive constants, we clearly have

$$d(A_1, A_2) \geq c_1/n$$

and

$$\sum_{u \in U} \sum_{v \in \bar{U}} \lambda(u) p(u, v) \geq \frac{c_2}{n} \lambda(B).$$

Hence we may assume that $\lambda(B)$ is small, say $\lambda(B) < 1/n$.

Define a measure $\mu$ on $\mathcal{B}(M)$, as in Theorem 4.2, with $F = f_1 = 2^{-\varphi_1}$, where $\varphi_1$ is the maximal convex function on $M$ dominated by $\varphi$. Note that, for every $u \in V$, $\lambda(u)$ is within a constant factor of the $\mu$-measure of the unit cube centred at $u$. Hence, by Theorem 4.2,

$$\frac{\sum_{u \in U} \sum_{v \in \bar{U}} \lambda(u) p(u, v)}{\lambda(U)} \geq \frac{c_3}{n}\,\frac{\mu(M \setminus A_1 \cup A_2)}{\min\{\mu(A_1), \mu(A_2)\}} \geq n^{-7},$$

since, rather crudely, $\operatorname{diam} M = O(n^5)$ and $d(A_1, A_2) \geq c_1/n$. This completes the sketch of a proof of Theorem 4.1.

We have made no attempt to get a really fast algorithm: in particular, one could use an isoperimetric inequality closer to the problem at hand than Theorem 4.2. Using a more careful analysis, Dyer and Frieze [1991] showed that the running time of the algorithm above is $O^*(n^8)$.

We conclude with a few words about the latest results concerning FRPAS for computing the volume. Improving the earlier results, Kannan, Lovász and Simonovits [1995] proved the following theorem.

THEOREM 4.3. *Let $c > 0$ be a constant and let $K \subset \mathbb{R}^n$ be a convex body given by a separation oracle, with guarantee*

$$B^n \subset K \subset n^c B^n.$$

*There is a fully randomized polynomial approximation scheme that, given $\varepsilon, \eta > 0$, returns positive numbers $\underline{\mathrm{vol}}\, K$ and $\overline{\mathrm{vol}}\, K$ such that $\underline{\mathrm{vol}}\, K \leq (1 + \varepsilon)\overline{\mathrm{vol}}\, K$ and*

$$\underline{\mathrm{vol}}\, K \leq \mathrm{vol}\, K \leq \overline{\mathrm{vol}}\, K$$

*with probability at least $1 - \eta$. This algorithm uses*

$$O\left(\frac{n^5}{\varepsilon^2}\left(\log\frac{1}{\varepsilon}\right)^3\left(\log\frac{1}{\eta}\right)(\log n)^5\right) = O^*\left(n^5\right)$$

*calls to the oracle.*

The basis of their proof is, once again, a fast sampling algorithm, that is, a fast algorithm that generates $N$ random points $v_1, v_2, \ldots, v_N$ of $K$ with *almost uniform* and *almost independent* distributions. To be precise, if

$$B^n \subset K \subset dB^n \tag{4.2}$$

then we can achieve that

(a) the distribution of each $v_i$ is close to the uniform distribution in the total variation distance: for $U \in \mathcal{B}(K)$ we have

$$|\mathbb{P}(v_i \in U) - \mathrm{vol}\, U / \mathrm{vol}\, K| < \varepsilon;$$

(b) for $1 \leq i < j \leq N$ and $A, B \in \mathcal{B}(K)$ we have

$$|\mathbb{P}(v_i \in A,\, v_j \in B) - \mathbb{P}(V_i \in A)\mathbb{P}(V_j \in B)| < \varepsilon;$$

(c) the algorithm uses only $O^*(n^3 d^2 + N n^2 d^2)$ calls to the oracle.

The main innovation in finding such an algorithm is that instead of demanding (4.2), Kannan, Lovász and Simonovits [$\geq$ 1997] are satisfied with 'approximate sandwiching', that is, if $B^n$ is contained in $K$ and $d'B^n \cap K$ is *most of $K$*, provided $d'$ is small and $K$ can be 'turned' into such a position by a fast algorithm. Thus we want an efficient way of finding an affine transformation $T$ such that $B^n \subset TK$ and $d'B^n \cap TK$ is most of $TK$ for $d'$ fairly small.

It is easy to show [Pisier 1989; Ball 1997] that (4.2) cannot be guaranteed with $d < n$; and it is not known whether in polynomial time one can guarantee it with $d$ not much larger than $n$. However, if we demand only approximate sandwiching then we *can* find an affine transformation $T$ in $O^*(n^5)$ time and $d' = O(\sqrt{n}/\log(1/\varepsilon))$. After much work, this leads to a volume algorithm with $O^*(n)$ calls to the oracle.

Finally, having emphasized how surprising it is that there are fully randomized polynomial time algorithms to approximate the volume of a convex body, let us note that there seems to be no nontrivial *lower bound* on the speed of such an algorithm. For example, it is not impossible that there are algorithms running in time $O^*(n^2)$.

# References

[Aldous 1987] D. Aldous, "On the Markov chain simulation method for uniform combinatorial distributions and simulated annealing", *Probab. Engrg. Inform. Sci.* **1** (1987), 33–46.

[Alon 1986] N. Alon, "Eigenvalues and expanders", *Combinatorica* **6** (1986), 86–96.

[Alon and Milman 1985] N. Alon and V. D. Milman, "$\lambda_1$, isoperimetric inequalities for graphs and superconcentrators", *J. Combin. Theory Ser. B* **38** (1985), 73–88.

[Applegate and Kannan 1991] D. L. Applegate and R. Kannan, "Sampling and integration of near log-concave functions", pp. 156–163 in *Proc. 23rd Annual ACM Symposium on Theory of Computing* (New Orleans, 1991), ACM, New York, 1991.

[Ball 1986] K. M. Ball, "Cube slicing in $\mathbb{R}^n$", *Proc. Amer. Math. Soc.* **97** (1986), 465–473.

[Ball 1997] K. M. Ball, "An elementary introduction to modern convex geometry", pp. 1–58 in *Flavors of Geometry*, edited by S. Levy, MSRI Publications **31**, Cambridge U. Press, New York, 1997.

[Ball and Pajor 1990] K. M. Ball and A. Pajor, "Convex bodies with few faces", *Proc. Amer. Math. Soc.* **110** (1990), 225–231.

[Bárány and Füredi 1986] I. Bárány and Z. Füredi, "Computing the volume is difficult", pp. 442–447 in *Proc. 18th Annual ACM Symposium on Theory of Computing* (Berkeley, 1986), ACM, New York, 1986.

[Bárány and Füredi 1987] I. Bárány and Z. Füredi, "Computing the volume is difficult", *Discrete Comput. Geom.* **2** (1987), 319–326.

[Bárány and Füredi 1988] I. Bárány and Z. Füredi, "Approximation of the sphere by polytopes having few vertices", *Proc. Amer. Math. Soc.* **102** (1988), 651–659.

[Bollobás 1979] B. Bollobás, *Graph theory: an introductory course*, Graduate Texts in Math. **63**, Springer, Heidelberg, 1979.

[Bombieri and Vaaler 1983] E. Bombieri and J. D. Vaaler, "On Siegel's lemma", *Invent. Math.* **73** (1983), 11–32.

[Bourgain et al. 1989] J. Bourgain, J. Lindenstrauss, and V. D. Milman, "Approximation of zonoids by zonotopes", *Acta Math.* **162** (1989), 73–141.

[Bourgain and Milman 1985] J. Bourgain and V. D. Milman, "Sections euclidiennes et volume des corps symétriques convexes dans $\mathbb{R}^n$", *C. R. Acad. Sci. Paris Sér. I* **300** (1985), 435–437.

[Carathéodory 1907] C. Carathéodory, "Über den Variabilitätsbereich der Koeffizienten von Potenzreihen, die gegebene Werte nicht annehmen", *Math. Ann.* **64** (1907), 95–115.

[Carl 1985] B. Carl, "Inequalities of Bernstein–Jackson-type and the degree of compactness of operators in Banach spaces", *Ann. Inst. Fourier (Grenoble)* **35**:3 (1985), 79–118.

[Carl and Pajor 1988] B. Carl and A. Pajor, "Gel'fand numbers of operators with values in a Hilbert space", *Invent. Math.* **94** (1988), 479–504.

[Cheeger 1970] J. Cheeger, "A lower bound for the smallest eigenvalue of the Laplacian", pp. 195–199 in *Problems in analysis* (papers dedicated to Salomon Bochner, 1969), edited by R. C. Gunning, Princeton U. Press, Princeton, 1970.

[Chung 1996] F. R. K. Chung, *Spectral graph theory*, CBMS Lecture Notes, Amer. Math. Soc., Providence, 1996.

[Dyer and Frieze 1988] M. E. Dyer and A. M. Frieze, "On the complexity of computing the volume of a polyhedron", *SIAM J. Comput.* **17** (1988), 967–974.

[Dyer and Frieze 1991] M. E. Dyer and A. M. Frieze, "Computing the volume of convex bodies: a case where randomness provably helps", pp. 123–169 in *Probabilistic combinatorics and its applications* (San Francisco, 1991), edited by B. Bollobás, Proc. Symp. Applied Math. **44**, Amer. Math. Soc., Providence, 1991.

[Dyer et al. 1991] M. E. Dyer, A. M. Frieze, and R. Kannan, "A random polynomial time algorithm for approximating the volume of convex bodies", *J. ACM* **38** (1991), 1–17.

[Eckhoff 1993] J. Eckhoff, "Helly, Radon, and Carathéodory type theorems", pp. 389–448 in *Handbook of convex geometry A*, edited by P. M. Gruber and J. M. Wills, North-Holland, Amsterdam, 1993.

[Elekes 1986] G. Elekes, "A geometric inequality and the complexity of measuring the volume", *Discrete Comput. Geom.* **1** (1986), 289–292.

[Fejes Tóth 1964] L. Fejes Tóth, *Regular Figures*, Monographs in Pure and Applied Math. **48**, MacMillan, New York, 1964.

[Figiel and Johnson 1980] T. Figiel and W. B. Johnson, "Large subspaces of $l_\infty^n$ and estimates of the Gordon–Lewis constants", *Israel J. Math.* **37** (1980), 92–112.

[Gluskin 1988] E. D. Gluskin, "Extremal properties of rectangular parallelipipeds and their applications to the geometry of Banach spaces", *Mat. Sb. (N. S.)* **136** (1988), 85–95. In Russian.

[Gordon et al. ≥ 1997] Y. Gordon, S. Reisner, and C. Schütt, "Umbrellas and polytopal approximation of the Euclidean ball". To appear.

[Grötschel et al. 1988] M. Grötschel, L. Lovász, and A. Schrijver, *Geometric algorithms and combinatorial optimization*, Algorithms and combinatorics **2**, Springer, Berlin, 1988.

[Hensley 1979]  D. Hensley, "Slicing the cube in $\mathbb{R}^n$ and probability (bounds for the measure of a central cube slice in $\mathbb{R}^n$ by probability methods)", *Proc. Amer. Math. Soc.* **73** (1979), 95–100.

[Jerrum and Sinclair 1989]  M. R. Jerrum and A. J. Sinclair, "Approximating the permanent", *SIAM J. Comput.* **18** (1989), 1149–1178.

[John 1948]  F. John, "Extremum problems with inequalities as subsidiary conditions", pp. 187–204 in *Studies and essays presented to R. Courant on his 60th birthday* (Jan. 8, 1948), Interscience, New York, 1948.

[Kemeny and Snell 1976]  J. G. Kemeny and J. L. Snell, *Finite Markov chains*, Undergraduate Texts in Mathematics, Springer, New York, 1976. Reprinting of the 1960 original.

[Khachiyan 1988]  L. G. Khachiyan, "On the complexity of computing the volume of a polytope", *Izvestia Akad. Nauk SSSR Tekhn. Kibernet.* **3** (1988), 216–217.

[Khachiyan 1989]  L. G. Khachiyan, "The problem of computing the volume of polytopes is NP-hard", *Uspekhi Mat. Nauk* **44** (1989), 179–180. In Russian; translation in *Russian Math. Surveys* **44** (1989), no. 3, 199–200.

[Khachiyan 1993]  L. G. Khachiyan, "Complexity of polytope volume computation", pp. 91–101 in *New trends in discrete and computational geometry*, edited by J. Pach, Algorithms and Combinatorics **10**, Springer, 1993.

[Lawrence 1991]  J. Lawrence, "Polytope volume computation", *Math. Comp.* **57**:195 (1991), 259–271.

[Lenstra 1983]  H. W. Lenstra, "Integer programming with a fixed number of variables", *Math. Oper. Res.* **8** (1983), 538–548.

[Lovász 1986]  L. Lovász, *An algorithmic theory of numbers, graphs and convexity*, CBMS–NSF Regional Conference Series in Applied Mathematics **50**, SIAM, Philadelphia, 1986.

[Lovász and Simonovits 1990]  L. Lovász and M. Simonovits, "The mixing rate of Markov chains, an isoperimetric inequality, and computing the volume", pp. 346–354 in *31st Annual Symposium on Foundations of Computer Science* (St. Louis, MO, 1990), IEEE Comput. Soc. Press, Los Alamitos, CA, 1990.

[Lovász and Simonovits 1992]  L. Lovász and M. Simonovits, "On the randomized complexity of volume and diameter", pp. 482–491 in *33st Annual Symposium on Foundations of Computer Science* (Pittsburgh, 1992), IEEE Comput. Soc. Press, Los Alamitos, CA, 1992.

[Lovász and Simonovits 1993]  L. Lovász and M. Simonovits, "Random walks in a convex body and an improved volume algorithm", *Random Structures and Algorithms* **4** (1993), 359–412.

[Mihail 1989]  M. Mihail, "Conductance and convergence of Markov chains—a combinatorial treatment of expanders", pp. 526–531 in *30th Annual Symposium on Foundations of Computer Science* (Research Triangle Park, NC, 1989), IEEE Comput. Soc. Press, Los Alamitos, CA, 1989.

[Payne and Weinberger 1960]  L. E. Payne and H. F. Weinberger, "An optimal Poincaré inequality for convex domains", *Arch. Rational Mech. Anal.* **5** (1960), 286–292.

[Pisier 1989]   G. Pisier, *The volume of convex bodies and Banach space geometry*, Cambridge Tracts in Mathematics, 94, Cambridge University Press, Cambridge, 1989.

[Ravi Kannan and Simonovits ≥ 1997]   L. L. Ravi Kannan and M. Simonovits, "Random walks and an $O^*(n^5)$ volume algorithm for convex bodies". to appear.

[Santaló 1949]   L. Santaló, "Un invariante afin para los cuerpos convexos del espacio de $n$ dimensiones", 1949.

[Sinclair and Jerrum 1989]   A. J. Sinclair and M. R. Jerrum, "Approximate counting, uniform generation and rapidly mixing Markov chains", *Inform. and Comput.* **82** (1989), 93–133.

[Vaaler 1979]   J. D. Vaaler, "A geometric inequality with applications to linear forms", *Pacific J. Math.* **83** (1979), 543–553.

[Vazirani 1991]   U. Vazirani, "Rapidly mixing Markov chains", pp. 99–121 in *Probabilistic Combinatorics and Its Applications* (San Francisco, 1991), edited by B. Bollobás, Proc. Symp. Applied Math. **44**, Amer. Math. Soc., Providence, 1991.

# Index

BÉLA BOLLOBÁS
DEPARTMENT OF MATHEMATICS
UNIVERSITY OF MEMPHIS
INSTITUTE FOR ADVANCED STUDY
  bollobab@ibex.msci.memphis.edu

Flavors of Geometry
MSRI Publications
Volume 31, 1997

# Cumulative Index

Printed in the United States
By Bookmasters